# Essentials of Enzymology

# Essentials of Enzymology

Colby Smith

Larsen & Keller
www.larsen-keller.com

Essentials of Enzymology
Colby Smith
ISBN: 978-1-64172-446-3 (Hardback)

☐ Larsen & Keller

Published by Larsen and Keller Education,
5 Penn Plaza,
19th Floor,
New York, NY 10001, USA

**Cataloging-in-Publication Data**

Essentials of enzymology / Colby Smith.
    p. cm.
Includes bibliographical references and index.
ISBN 978-1-64172-446-3
1. Enzymology. 2. Enzymes. 3. Biochemistry. I. Smith, Colby.
QP601 .E87 2020
572.7--dc23

For more information regarding Larsen and Keller Education and its products, please visit the publisher's website www.larsen-keller.com

# Table of Contents

# Preface

The macromolecular biological catalysts which accelerate chemical reactions are known as enzymes. They are an integral part of most of the metabolic processes in cells. They are involved in the conversion of substrates into different molecules known as products. The study of enzymes is known as enzymology. It deals with the study of their kinetics, structure and function. Enzyme kinetics is a sub-field of enzymology which seeks to study the processes through which substrates are bound by enzymes and turned into products. Some of the other functions of enzymes which are studied within this field are signal transduction and cell regulation. Different approaches, evaluations and methodologies and advanced studies on enzymes have been included in this textbook. While understanding the long-term perspectives of the topics, the book makes an effort in highlighting their impact as a modern tool for the growth of the discipline. The coherent flow of topics, student-friendly language and extensive use of examples make this book an invaluable source of knowledge.

Given below is the chapter wise description of the book:

Chapter 1- The macromolecular biological catalysts which accelerate chemical reactions are known as enzymes. Similar to other catalysts, enzymes lower the activation energy, thereby increasing the reaction rate. Some of the different kinds of enzymes are oxidoreductases, transferases, hydrolases, lyases, isomerases and hexokinase. This is an introductory chapter which will provide a brief introduction to all these significant aspects of enzymes.

Chapter 2- The study of chemical reactions which are catalyzed by enzymes is known as enzyme reaction kinetics. These reactions can be affected by molecules known as enzyme inhibitors and activators. The enzyme inhibitors reduce or abolish the activity of enzymes while enzyme activators increase the catalytic rate of enzymes. This chapter discusses in detail the theories and concepts related to enzyme reaction kinetics and enzyme preparation.

Chapter 3- The molecules on which the enzymes act are known as substrates. The region where substrate molecules bind and undergo a chemical reaction is known as an active site. In order for the enzyme to function as a catalyst, a non-protein chemical compound or a metallic ion known as an enzyme cofactor is required. This chapter has been carefully written to provide an easy understanding of the varied facets of active sites and enzyme cofactors.

Chapter 4- A majority of enzymes are produced by the fungi. They are integral as a source of biological diversity as well for the production of industrial enzyme products. Some of the important fungal enzymes are fungal ligninolytic enzymes, proteolytic enzyme, fungal phytases and fungal amylase. The chapter closely examines the key concepts of these fungal enzymes to provide an extensive understanding of the subject.

Chapter 5- Enzymes which are used for commercial purposes in various industries are known as industrial enzymes. Pharmaceuticals, biofuels, consumer products and chemical production are a few industries which make use of these enzymes. The chapter closely examines these industrial enzymes to provide an extensive understanding of the subject.

Indeed, my job was extremely crucial and challenging as I had to ensure that every chapter is informative and structured in a student-friendly manner. I am thankful for the support provided by my family and colleagues during the completion of this book.

**Colby Smith**

# An Introduction to Enzymes

The macromolecular biological catalysts which accelerate chemical reactions are known as enzymes. Similar to other catalysts, enzymes lower the activation energy, thereby increasing the reaction rate. Some of the different kinds of enzymes are oxidoreductases, transferases, hydrolases, lyases, isomerases and hexokinase. This is an introductory chapter which will provide a brief introduction to all these significant aspects of enzymes.

Enzymes are proteins that act as catalysts within living cells. Catalysts increase the rate at which chemical reactions occur without being consumed or permanently altered themselves. A chemical reaction is a process that converts one or more substances (known as reagents, reactants, or substrates) to another type of substance (the product). As a catalyst, an enzyme can facilitate the same chemical reaction over and over again.

## Structure and Function

Like all proteins, enzymes are composed of one or more long chains of interconnected amino acids. Each enzyme possesses a unique sequence of amino acids that causes it to fold into a characteristic shape. An enzyme's amino acid sequence is determined by a specific gene in the cell's nucleus. This ensures that each copy of the enzyme is the same as all others.

On the surface of each enzyme is a special cleft called the active site, which provides a place where reagents can 'meet' and interact. Much like a lock and its key, an enzyme's active site will only accommodate certain reagents, and only one type of chemical reaction can be catalyzed by a given enzyme.

For example, during the manufacture of hemoglobin (the oxygen-carrying pigment in your red blood cells), a single atom of iron must be inserted into the center of the molecule to make it functional. An enzyme called ferrochelatase brings the reagents (iron and the empty molecule) together, catalyzes their union, and releases an iron-containing molecule. This is the only reaction catalyzed by ferrochelatase. Keep in mind that enzymes can combine reagents (as in the synthesis of hemoglobin), they can split a single reagent into multiple products, or they can simply transform a single reagent into a single product that looks different from the original reagent.

When reagents enter an enzyme's active site, the enzyme undergoes a temporary change in shape that encourages interaction between the reagents. Upon completion of the chemical reaction, a specific product is released from the active site, the enzyme resumes its original conformation, and the reaction can begin again with new reagents.

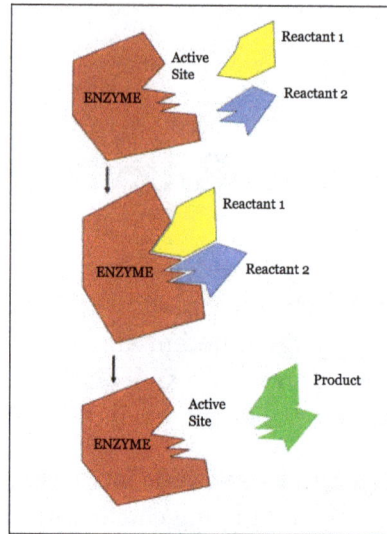

Many enzymes are incorporated into metabolic pathways. A metabolic pathway is a series of chemical reactions that transform one or more reagents into an end-product that's needed by the cell. The enzymes in a metabolic pathway-much like people passing a pail of water along a bucket brigade-move a reagent along until the end-product is produced. A metabolic pathway can be quite short, or it can have many steps and multiple enzymes. The metabolic pathway that converts tryptophan (an amino acid found in dietary protein) to serotonin (a chemical that's necessary for normal brain function) is only two steps long.

In order to function, many enzymes require the help of cofactors or coenzymes. Cofactors are often metal ions, such as zinc, copper, iron, or magnesium. Magnesium, one of the most common cofactors, activates hundreds of enzymes, including those that manufacture DNA and many that help metabolize carbohydrates.

Many coenzymes are derived from vitamins. In fact, one of the main reasons you need vitamins in your diet is to supply the raw material for essential coenzymes. For example, vitamin C is needed by the enzyme that produces collagen and builds healthy skin, a coenzyme derived from vitamin B12 is necessary for synthesizing the insulation around your nerve cells, and a vitamin B6-based coenzyme is vital for producing serotonin.

Coenzymes and cofactors bind to the active sites of enzymes, and they participate in catalysis, but they are not generally considered reagents, nor do they become part of the product(s) of the reaction. In many cases, cofactors and coenzymes function as intermediate carriers of electrons, specific atoms, or functional groups that are transferred during the overall reaction.

A large protein enzyme molecule is composed of one or more amino acid chains called polypeptide chains. The amino acid sequence determines the characteristic folding patterns of the protein's structure, which is essential to enzyme specificity. If the enzyme is subjected to changes, such as fluctuations in temperature or pH, the protein structure may lose its integrity (denature) and its enzymatic ability. Denaturation is sometimes, but not always, reversible.

Bound to some enzymes is an additional chemical component called a cofactor, which is a direct participant in the catalytic event and thus is required for enzymatic activity. A cofactor may be either a coenzyme—an organic molecule, such as a vitamin—or an inorganic metal ion; some enzymes require both. A cofactor may be either tightly or loosely bound to the enzyme. If tightly connected, the cofactor is referred to as a prosthetic group.

## Nomenclature

An enzyme will interact with only one type of substance or group of substances, called the substrate, to catalyze a certain kind of reaction. Because of this specificity, enzymes often have been named by adding the suffix "-ase" to the substrate's name (as in urease, which catalyzes the breakdown of urea). Not all enzymes have been named in this manner, however, and to ease the confusion surrounding enzyme nomenclature, a classification system has been developed based on the type of reaction the enzyme catalyzes. There are six principal categories and their reactions: (1) oxidoreductases, which are involved in electron transfer; (2) transferases, which transfer a chemical group from one substance to another; (3) hydrolases, which cleave the substrate by uptake of a water molecule (hydrolysis); (4) lyases, which form double bonds by adding or removing a chemical group; (5) isomerases, which transfer a group within a molecule to form an isomer; and (6) ligases, or synthetases, which couple the formation of various chemical bonds to the breakdown of a pyrophosphate bond in adenosine triphosphate or a similar nucleotide.

## Mechanism of Enzyme Action

In most chemical reactions, an energy barrier exists that must be overcome for the reaction to occur. This barrier prevents complex molecules such as proteins and nucleic acids from spontaneously degrading, and so is necessary for the preservation of life. When metabolic changes are required in a cell, however, certain of these complex molecules must be broken down, and this energy barrier must be surmounted. Heat could provide the additional needed energy (called activation energy), but the rise in temperature would kill the cell. The alternative is to lower the activation energy level through the use of a catalyst. This is the role that enzymes play. They react with the substrate to form an intermediate complex—a "transition state"—that requires less energy for the reaction to proceed. The unstable intermediate compound quickly breaks down to form reaction products, and the unchanged enzyme is free to react with other substrate molecules.

Only a certain region of the enzyme, called the active site, binds to the substrate. The active site is a groove or pocket formed by the folding pattern of the protein. This three-dimensional structure, together with the chemical and electrical properties of the amino acids and cofactors within the active site, permits only a particular substrate to bind to the site, thus determining the enzyme's specificity.

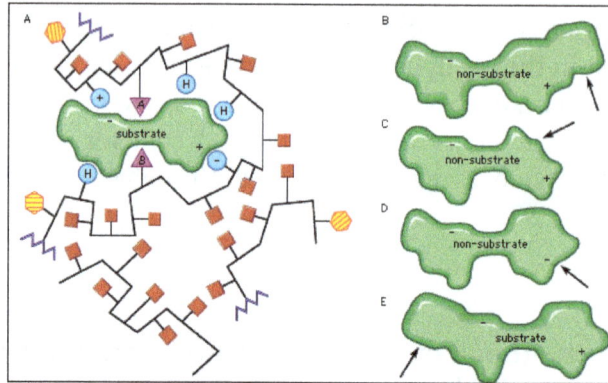

Enzyme; active site the active site of an enzyme is a groove or pocket that binds a specific substrate.

Enzyme synthesis and activity also are influenced by genetic control and distribution in a cell. Some enzymes are not produced by certain cells, and others are formed only when required. Enzymes are not always found uniformly within a cell; often they are compartmentalized in the nucleus, on the cell membrane, or in subcellular structures. The rates of enzyme synthesis and activity are further influenced by hormones, neurosecretions, and other chemicals that affect the cell's internal environment.

## Properties and Mechanism

In some cases, the increase in the rate of an enzyme-catalyzed reaction versus the uncatalyzed rate is a millionfold. As soon as one reaction has been catalyzed, the enzyme is available for another round of catalysis —a phenomenon known as turnover. Enzymes operate near physiological temperature and pH, and they are also highly specific in their actions—for example, an enzyme called hexokinase will place a phosphate group only onto the sixth carbon of a D-glucose molecule. The enzyme has no activity toward L-glucose, and reduced activity toward other D-sugars. Enzymes can also be regulated so that they are switched on only when they are needed by the cell. They may consist of a single polypeptide chain of amino acids (RNAse contains 124 amino acids), or they may require an additional chemical called a coenzyme. Many of the vitamins as well as several metals act as coenzymes.

A study of enzyme catalysis is a study of kinetics, which asks the question "how fast?" However, enzymes cannot alter the outcome or direction of a reaction. For instance, if one were to add a small amount of sodium chloride to a large volume of water, we know that the end result will be that the salt will dissolve in the water. However, the time dissolution takes depends on a number of factors: What is the temperature?; Is it being stirred? This is kinetics. We also know that a swinging pendulum will eventually come to rest at its equilibrium point, which in this case is its pointing straight down toward the center of Earth. Kinetics describes only the time it takes to reach that point. Enzymes cannot alter the equilibrium point of a reaction, only the time it takes to get there.

To proceed to products, reactants must come together with sufficient energy to overcome an energy barrier known as the energy of activation. The apex of this barrier represents the transition state between reactants and products. Enzymes act to lower the energy of activation by stabilizing (lowering the energy of) the transition state.

Mechanistically, an enzyme will bind the reactant, called the substrate, at a very specific site on the enzyme known as the active site. This resulting enzyme–substrate complex (ES), described as a lock-and-key mechanism, involves weak binding and often some structural changes—known as induced fit—that assist in stabilizing the transition state. In the unique microenvironment of the active site, substrate can rapidly be converted to product resulting in an enzyme product (EP) complex that then dissociates to release product.

## RNA as an Enzyme

Although enzymes are considered to be proteins, enzyme activity has recently been found in ribonucleic acid (RNA) in certain organisms. These "ribozymes" may yield clues to the origins of life on Earth. DNA needs enzymes to replicate, whereas enzymes need the instructions of DNA. This represents a "chicken-and-egg" question that has stumped researchers. Early life may have used RNA that was able to catalyze its own replication.

# Types of Enzymes

## Oxidoreductases

Oxidoreductases consist of a large class of enzymes catalyzing the transfer of electrons from an electron donor (reductant) to an electron acceptor (oxidant) molecule, generally taking nicotinamide adenine dinucleotide phosphate (NADP) or nicotinamide adenine dinucleotide (NAD) as cofactors. Since so many chemical and biochemical transformations comprise oxidation/reduction processes, it has long been an important goal in biotechnology to develop practical biocatalytic applications of oxidoreductases. During the past few years, significant breakthrough has been made in the development of oxidoreductase-based diagnostic tests and improved biosensors, and the design of innovative systems for the regeneration of essential coenzymes. Research on the construction of bioreactors for pollutants biodegradation and biomass processing, and the development of oxidoreductase-based approaches for synthesis of polymers and functionalized organic substrates have made great progress. Proper names of oxidore-ductases are in a form of "donor:acceptor oxidoreductase"; while in most cases "donor dehydrogenase" is much more common. Common names also sometimes appeared as "acceptor reductase", such as NAD+ reductase. "Donor oxidase" is a special case when O2 serves as the acceptor.

## Classification

Oxidoreuctases can be either oxidases or dehydrogenases. Oxidases are generally in-volved when molecular oxygen functions as an acceptor of hydrogen or electrons. How-ever, dehydrogenases work by oxidizing a substrate through transferring hydrogen to an acceptor that is either NAD/NADP or a flavin enzyme. Peroxidases, hydroxylases, oxygenases, and reductases also belong to oxidoreductases. Peroxidases are placed in peroxisomes, and could catalyze the reduction of hydrogen peroxide. Hydroxylases give hydroxyl groups to its substrates. Oxygenases could incorporate oxygen from mo-lecular oxygen into organic substrates. In most cases, reductases can act like oxidases, but catalyzing reductions.

Oxidoreductases are sorted as EC 1 in the EC number classification of enzymes and can be further classified into 22 subclasses.

| EC number | Description |
|-----------|-------------|
| EC 1.1 | Act on the CH-OH group of donors |
| EC 1.2 | Act on the aldehyde or oxo group of donors |
| EC 1.3 | Function on the CH-CH group of donors |
| EC 1.4 | Functioning on the CH-NH$_2$ group of donors |
| EC 1.5 | Act on CH-NH group of donors |
| EC 1.6 | Act on NADH or NADPH |
| EC 1.7 | Take other nitrogenous compounds as donors |
| EC 1.8 | Act on a sulfur group of donors |
| EC 1.9 | Act on a heme group of donors, respectively |
| EC 1.10 | Treating diphenols and related substances as donors |
| EC 1.11 | Act on peroxide as an acceptor (peroxidases) |
| EC 1.12 | Act on hydrogen as donors |
| EC 1.13 | Act on single donors with incorporation of molecular oxygen |
| EC 1.14 | Function on paired donors with incorporation of oxygen |
| EC 1.15 | Act on and act on superoxide radicals as acceptors |
| EC 1.16 | Oxidize metal ions |
| EC 1.17 | Take effect on CH or CH$_2$ groups |

| EC 1.18 | Apply iron-sulfur proteins as donors |
|---------|--------------------------------------|
| EC 1.19 | Take reduced flavodoxin as donors |
| EC 1.20 | Dispose phosphorus or arsenic in donors |
| EC 1.21 | Form a X-Y bond from X-H and Y-H bond |
| EC 1.97 | Some other oxidoreductases |

## Reactions

The catalyzed reactions are similar to the following reaction in Figure where A is the reductant and B is the oxidant. In biochemical reactions, the redox reactions are sometimes more difficult to observe, such as this reaction from glycolysis: Pi + glyceraldehyde-3-phosphate $NAD^+ \rightarrow NADH+H^+ +1,3-$ bisphosphoglycerate, where $NAD^+$ is the oxidant (electron acceptor), and glyceraldehyde-3-phosphate functions as reductant (electron donor).

$$A_{red} + B_{OX} \rightarrow A_{OX} + B_{red}$$

## Functions

Oxidoreductase enzymes play significant roles in both aerobic and anaerobic metabolism. They can be found in biological procedures like glycolysis, TCA cycle, oxidative phosphorylation, and amino acid metabolism. In glycolysis, the enzyme glyceraldehydes-3-phosphate dehydrogenase accelerates the reduction of $NAD^+$ to NADH. However, the re-oxidization of the generated NADH to $NAD^+$ occurs in the oxidative phosphorylation pathway in order to maintain the redox state of the cell. Additional NADH molecules are produced in the TCA cycle. The glycolysis product pyruvate takes part in the TCA cycle in a form of acetyl-CoA. During anaerobic glycolysis, the oxidation of NADH is accomplished through the reduction of pyruvate to lactate, which is then oxidized to pyruvate in muscle and liver cells. Moreover, the pyruvate is further oxidized in the TCA cycle. All twenty of the amino acids, except for leucine and lysine, can be degraded to intermediates in TCA cycle, which allows the carbon skeletons of the amino acids to be converted into oxaloacetate and subsequently into pyruvate. The gluconeogenic pathway can then exploit the formed pyruvate.

## Transferases

A transferase is any one of a class of enzymes that enact the transfer of specific functional groups (e.g. a methyl or glycosyl group) from one molecule (called the donor) to another (called the acceptor). They are involved in hundreds of different biochemical pathways throughout biology, and are integral to some of life's most important processes.

Transferases are involved in myriad reactions in the cell. Three examples of these re-actions are the activity of coenzyme A (CoA) transferase, which transfers thiol esters, the action of N-acetyltransferase, which is part of the pathway that metabolizes trypto-phan, and the regulation of pyruvate dehydrogenase (PDH), which converts pyruvate to acetyl CoA. Transferases are also utilized during translation. In this case, an amino acid chain is the functional group transferred by a peptidyl transferase. The transfer involves the removal of the growing amino acid chain from the tRNA molecule in the A-site of the ribosome and its subsequent addition to the amino acid attached to the tRNA in the P-site.

Mechanistically, an enzyme that catalyzed the following reaction would be a transfer-ase:

$$Xgroup + Y \xrightarrow{transferase} X + Ygroup$$

In the above reaction, X would be the donor, and Y would be the acceptor. "Group" would be the functional group transferred as a result of transferase activity. The donor is often a coenzyme.

## Nomenclature

Systematic names of transferases are constructed in the form of "donor:acceptor grouptransferase." For example, methylamine:L-glutamate N-methyltransferase would be the standard naming convention for the transferase methylamine-glutamate N-meth-yltransferase, where methylamine is the donor, L-glutamate is the acceptor, and meth-yltransferase is the EC category grouping. This same action by the transferase can be illustrated as follows:

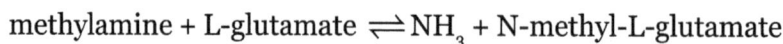

$$methylamine + L\text{-}glutamate \rightleftharpoons NH_3 + N\text{-}methyl\text{-}L\text{-}glutamate$$

However, other accepted names are more frequently used for transferases, and are of-ten formed as "acceptor grouptransferase" or "donor grouptransferase." For example, a DNA methyltransferase is a transferase that catalyzes the transfer of a methyl group to a DNA acceptor. In practice, many molecules are not referred to using this terminology due to more prevalent common names. For example, RNA Polymerase is the modern common name for what was formerly known as RNA nucleotidyltransferase, a kind of nucleotidyl transferase that transfers nucleotides to the 3' end of a growing RNA strand. In the EC system of classification, the accepted name for RNA Polymerase is DNA-directed RNA polymerase.

## Classification

Described primarily based on the type of biochemical group transferred, transferas-es can be divided into ten categories (based on the EC Number classification). These categories comprise over 450 different unique enzymes. In the EC numbering system, transferases have been given a classification of EC2. Hydrogen is not considered a functional group when it comes to transferase targets; instead, hydrogen transfer is included under oxidoreductases, due to electron transfer considerations.

| Classification of transferases into subclasses | | |
|---|---|---|
| EC number | Examples | Group(s) transferred |
| EC 2.1 | methyltransferase and formyltransferase | single-carbon groups |
| EC 2.2 | transketolase and transaldolase | aldehyde or ketone groups |
| EC 2.3 | acyltransferase | acyl groups or groups that become alkyl groups during transfer |
| EC 2.4 | glycosyltransferase, hexosyltransferase, and pentosyltransferase | glycosyl groups, as well as hexoses and pentoses |
| EC 2.5 | riboflavin synthase and chlorophyll synthase | alkyl or aryl groups, other than methyl groups |
| EC 2.6 | transaminase, and oximinotransferase | nitrogenous groups |
| EC 2.7 | phosphotransferase, polymerase, and kinase | phosphorus-containing groups; subclasses are based on the acceptor (e.g. alcohol, carboxyl, etc.) |
| EC 2.8 | sulfurtransferase and sulfotransferase | sulfur-containing groups |
| EC 2.9 | selenotransferase | selenium-containing groups |
| EC 2.10 | molybdenumtransferase and tungstentransferase | molybdenum or tungsten |

## Reactions

## EC 2.1: Single Carbon Transferases

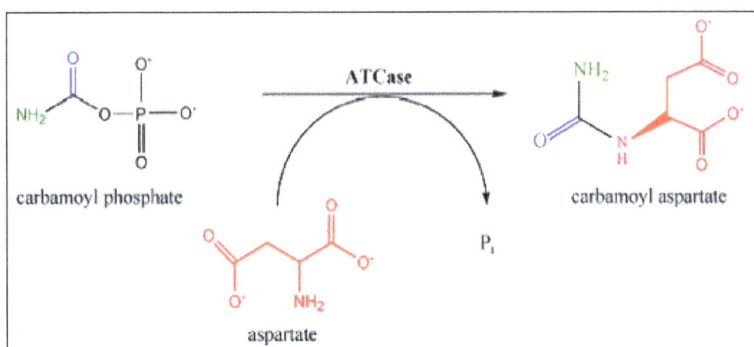

Reaction involving aspartate transcarbamylase.

EC 2.1 includes enzymes that transfer single-carbon groups. This category consists of transfers of methyl, hydroxymethyl, formyl, carboxy, carbamoyl, and amido groups. Carbamoyltransferases, as an example, transfer a carbamoyl group from one molecule to another. Carbamoyl groups follow the formula $NH_2CO$. In ATCase such a transfer is written as Carbamyl phosphate + L-aspertate $\rightarrow$ L-carbamyl aspartate + phosphate.

# EC 2.2: Aldehyde and Ketone Transferases

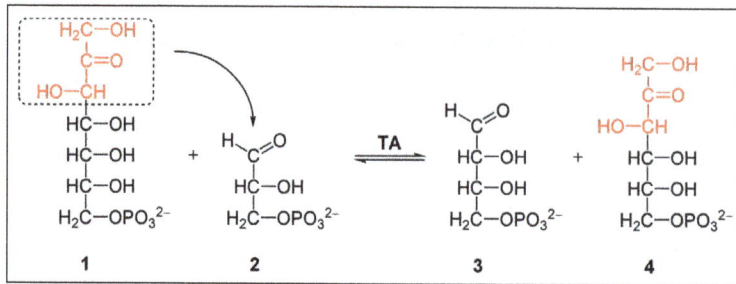

The reaction catalyzed by transaldolase.

Enzymes that transfer aldehyde or ketone groups and included in EC 2.2. This category consists of various transketolases and transaldolases. Transaldolase, the namesake of aldehyde transferases, is an important part of the pentose phosphate pathway. The reaction it catalyzes consists of a transfer of a dihydroxyacetone functional group to Glyceraldehyde 3-phosphate (also known as G3P). The reaction is as follows: sedoheptulose 7-phosphate + glyceraldehyde 3-phosphate $\rightleftharpoons$ erythrose 4-phosphate + fructose 6-phosphate.

# EC 2.3: Acyl Transferases

Transfer of acyl groups or acyl groups that become alkyl groups during the process of being transferred are key aspects of EC 2.3. Further, this category also differentiates between amino-acyl and non-amino-acyl groups. Peptidyl transferase is a ribozyme that facilitates formation of peptide bonds during translation. As an aminoacyltransferase, it catalyzes the transfer of a peptide to an aminoacyl-tRNA, following this reaction: peptidyl-tRNA$_A$ + aminoacyl-tRNA$_B$ $\rightleftharpoons$ tRNA$_A$ + peptidyl aminoacyl-tRNA$_B$.

# EC 2.4: Glycosyl, Hexosyl and Pentosyl Transferases

EC 2.4 includes enzymes that transfer glycosyl groups, as well as those that transfer hexose and pentose. Glycosyltransferase is a subcategory of EC 2.4 transferases that is involved in biosynthesis of disaccharides and polysaccharides through transfer of monosaccharides to other molecules. An example of a prominent glycosyltransferase is lactose synthase which is a dimer possessing two protein subunits. Its primary action is to produce lactose from glucose and UDP-galactose. This occurs via the following pathway: UDP-β-D-galactose + D-glucose $\rightleftharpoons$ UDP + lactose.

# EC 2.5: Alkyl and Aryl Transferases

EC 2.5 relates to enzymes that transfer alkyl or aryl groups, but does not include methyl groups. This is in contrast to functional groups that become alkyl groups when transferred, as those are included in EC 2.3. EC 2.5 currently only possesses one sub-class: Alkyl and aryl transferases. Cysteine synthase, for example, catalyzes the formation of acetic acids and cysteine from $O_3$-acetyl-L-serine and hydrogen sulfide: $O_3$-acetyl-L-serine + $H_2S \rightleftharpoons$ L-cysteine + acetate.

## EC 2.6: Nitrogenous Transferases

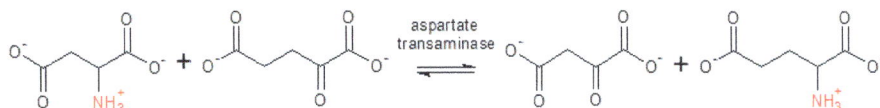

Aspartate aminotransferase can act on several different amino acids.

The grouping consistent with transfer of nitrogenous groups is EC 2.6. This includes enzymes like transaminase (also known as "aminotransferase"), and a very small number of oximinotransferases and other nitrogen group transferring enzymes. EC 2.6 previously included amidinotransferase but it has since been reclassified as a subcategory of EC 2.1 (single-carbon transferring enzymes). In the case of aspartate transaminase, which can act on tyrosine, phenylalanine, and tryptophan, it reversibly transfers an amino group from one molecule to the other.

The reaction, for example, follows the following order: L-aspartate +2-oxoglutarate oxaloacetate + L-glutamate.

## EC 2.7: Phosphorus Transferases

While EC 2.7 includes enzymes that transfer phosphorus-containing groups, it also includes nuclotidyl transferases as well. Sub-category phosphotransferase is divided up in categories based on the type of group that accepts the transfer. Groups that are classified as phosphate acceptors include: alcohols, carboxy groups, nitrogenous groups, and phosphate groups. Further constituents of this subclass of transferases are various kinases. A prominent kinase is cyclin-dependent kinase (or CDK), which comprises a sub-family of protein kinases. As their name implies, CDKs are heavily dependent on specific cyclin molecules for activation. Once combined, the CDK-cyclin complex is capable of enacting its function within the cell cycle.

The reaction catalyzed by CDK is as follows: ATP + a target protein → ADP + a phosphoprotein.

## EC 2.8: Sulfur Transferases

Transfer of sulfur-containing groups is covered by EC 2.8 and is subdivided into the subcategories of sulfurtransferases, sulfotransferases, and CoA-transferases, as well as enzymes that transfer alkylthio groups. A specific group of sulfotransferases are those that use PAPS as a sulfate group donor. Within this group is alcohol sulfotransferase which has a broad targeting capacity. Due to this, alcohol sulfotransferase is also known by several other names including "hydroxysteroid sulfotransferase," "steroid sulfokinase," and "estrogen sulfotransferase." Decreases in its activity has been linked to human liver disease. This transferase acts via the following reaction: 3'-phosphoadenylyl sulfate + an alcohol $\rightleftharpoons$ adenosine 3', 5' bisphosphate + an alkyl sulfate.

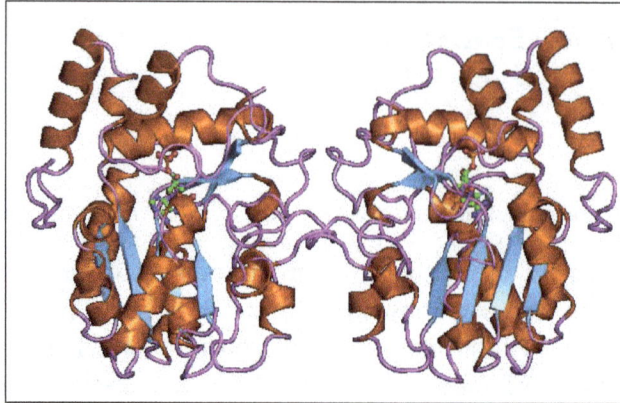

Ribbon diagram of a variant structure of estrogen sulfotransferase (PDB 1aqy EBI).

## EC 2.9: Selenium Transferases

EC 2.9 includes enzymes that transfer selenium-containing groups. This category only contains two transferases, and thus is one of the smallest categories of transferase. Selenocysteine synthase, which was first added to the classification system in 1999, converts seryl-tRNA(Sec UCA) into selenocysteyl-tRNA(Sec UCA).

## EC 2.10: Metal Transferases

The category of EC 2.10 includes enzymes that transfer molybdenum or tungsten-containing groups. However, as of 2011, only one enzyme has been added: molybdopterin molybdotransferase. This enzyme is a component of MoCo biosynthesis in *Escherichia coli*. The reaction it catalyzes is as follows: adenylyl-molybdopterin + molybdate $\rightarrow$ molybdenum cofactor + AMP.

## Role in Histo-blood Group

The A and B transferases are the foundation of the human ABO blood group system. Both A and B transferases are glycosyltransferases, meaning they transfer a sugar molecule onto an H-antigen. This allows H-antigen to synthesize the glycoprotein and glycolipid conjugates that are known as the A/B antigens. The full name of A transferase is alpha 1-3-N-acetylgalactosaminyltransferase and its function in the cell is to add N-acetylgalactosamine to H-antigen, creating A-antigen. The full name of B transferase is alpha 1-3-galactosyltransferase, and its function in the cell is to add a galactose molecule to H-antigen, creating B-antigen.

It is possible for *Homo sapiens* to have any of four different blood types: Type A (express A antigens), Type B (express B antigens), Type AB (express both A and B antigens) and Type O (express neither A nor B antigens). The gene for A and B transferases is located on chromosome 9. The gene contains seven exons and six introns and the gene itself is over 18kb long. The alleles for A and B transferases are extremely similar. The resulting enzymes only differ in 4 amino acid residues. The differing residues are located at positions 176, 235, 266, and 268 in the enzymes.

## Deficiencies

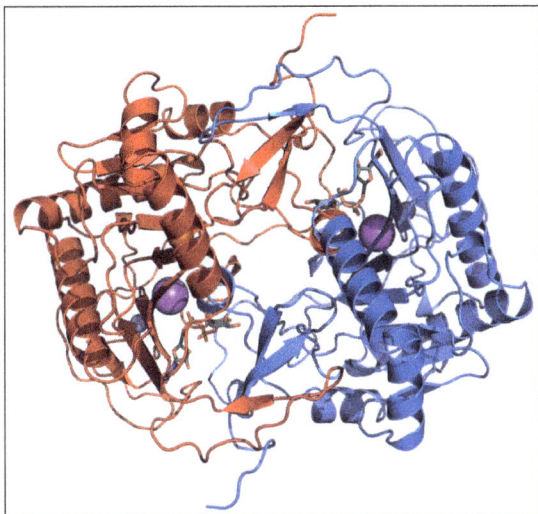

A deficiency of this transferase, E. coli galactose-1-phosphate uridyltransferase is a known cause of galactosemia.

Transferase deficiencies are at the root of many common illnesses. The most common result of a transferase deficiency is a buildup of a cellular product.

## SCOT Deficiency

Succinyl-CoA:3-ketoacid CoA transferase deficiency (or SCOT deficiency) leads to a buildup of ketones. Ketones are created upon the breakdown of fats in the body and are an important energy source. Inability to utilize ketones leads to intermittent ketoacidosis, which usually first manifests during infancy. Disease sufferers experience nausea, vomiting, inability to feed, and breathing difficulties. In extreme cases, ketoacidosis can lead to coma and death. The deficiency is caused by mutation in the gene OXCT1. Treatments mostly rely on controlling the diet of the patient.

## CPT-II Deficiency

Carnitine palmitoyltransferase II deficiency (also known as CPT-II deficiency) leads to an excess long chain fatty acids, as the body lacks the ability to transport fatty acids into the mitochondria to be processed as a fuel source. The disease is caused by a defect in the gene CPT2. This deficiency will present in patients in one of three ways: lethal neonatal, severe infantile hepatocardiomuscular, and myopathic form. The myopathic is the least severe form of the deficiency and can manifest at any point in the lifespan of the patient. The other two forms appear in infancy. Common symptoms of the lethal neonatal form and the severe infantile forms are liver failure, heart problems, seizures and death. The myopathic form is characterized by muscle pain and weakness following vigorous exercise. Treatment generally includes dietary modifications and carnitine supplements.

## Galactosemia

Galactosemia results from an inability to process galactose, a simple sugar. This deficiency occurs when the gene for galactose-1-phosphate uridylyltransferase (GALT) has any number of mutations, leading to a deficiency in the amount of GALT produced. There are two forms of Galactosemia: classic and Duarte. Duarte galactosemia is generally less severe than classic galactosemia and is caused by a deficiency of galactokinase. Galactosemia renders infants unable to process the sugars in breast milk, which leads to vomiting and anorexia within days of birth. Most symptoms of the disease are caused by a buildup of galactose-1-phosphate in the body. Common symptoms include liver failure, sepsis, failure to grow, and mental impairment, among others. Buildup of a second toxic substance, galactitol, occurs in the lenses of the eyes, causing cataracts. Currently, the only available treatment is early diagnosis followed by adherence to a diet devoid of lactose, and prescription of antibiotics for infections that may develop.

## Choline Acetyltransferase Deficiencies

Choline acetyltransferase (also known as ChAT or CAT) is an important enzyme which produces the neurotransmitter acetylcholine. Acetylcholine is involved in many neuropsychic functions such as memory, attention, sleep and arousal. The enzyme is globular in shape and consists of a single amino acid chain. ChAT functions to transfer an acetyl group from acetyl co-enzyme A to choline in the synapses of nerve cells and exists in two forms: soluble and membrane bound. The ChAT gene is located on chromosome 10.

## Alzheimer's Disease

Decreased expression of ChAT is one of the hallmarks of Alzheimer's disease. Patients with Alzheimer's disease show a 30 to 90% reduction in activity in several regions of the brain, including the temporal lobe, the parietal lobe and the frontal lobe. However, ChAT deficiency is not believed to be the main cause of this disease.

## Amyotrophic Lateral Sclerosis (ALS or Lou Gehrig's Disease)

Patients with ALS show a marked decrease in ChAT activity in motor neurons in the spinal cord and brain. Low levels of ChAT activity are an early indication of the disease and are detectable long before motor neurons begin to die. This can even be detected before the patient is symptomatic.

## Huntington's Disease

Patients with Huntington's also show a marked decrease in ChAT production. Though the specific cause of the reduced production is not clear, it is believed that the death of medium-sized motor neurons with spiny dendrites leads to the lower levels of ChAT production.

## Schizophrenia

Patients with Schizophrenia also exhibit decreased levels of ChAT, localized to the mesopontine tegment of the brain and the nucleus accumbens, which is believed to correlate with the decreased cognitive functioning experienced by these patients.

## Sudden Infant Death Syndrome (SIDS)

Recent studies have shown that SIDS infants show decreased levels of ChAT in both the hypothalamus and the striatum. SIDS infants also display fewer neurons capable of producing ChAT in the vagus system. These defects in the medulla could lead to an inability to control essential autonomic functions such as the cardiovascular and respiratory systems.

## Congenital Myasthenic Syndrome (CMS)

CMS is a family of diseases that are characterized by defects in neuromuscular transmission which leads to recurrent bouts of apnea (inability to breathe) that can be fatal. ChAT deficiency is implicated in myasthenia syndromes where the transition problem occurs presynaptically. These syndromes are characterized by the patients' inability to resynthesize acetylcholine.

## Uses in Biotechnology

## Terminal Transferases

Terminal transferases are transferases that can be used to label DNA or to produce plasmid vectors. It accomplishes both of these tasks by adding deoxynucleotides in the form of a template to the downstream end or 3' end of an existing DNA molecule. Terminal transferase is one of the few DNA polymerases that can function without an RNA primer.

## Glutathione Transferases

The family of glutathione transferases (GST) is extremely diverse, and therefore can be used for a number of biotechnological purposes. Plants use glutathione transferases as a means to segregate toxic metals from the rest of the cell. These glutathione transferases can be used to create biosensors to detect contaminants such as herbicides and insecticides. Glutathione transferases are also used in transgenic plants to increase resistance to both biotic and abiotic stress. Glutathione transferases are currently being explored as targets for anti-cancer medications due to their role in drug resistance. Further, glutathione transferase genes have been investigated due to their ability to prevent oxidative damage and have shown improved resistance in transgenic cultigens.

## Rubber Transferases

Currently the only available commercial source of natural rubber is the Hevea plant (Hevea brasiliensis). Natural rubber is superior to synthetic rubber in a number of commercial uses. Efforts are being made to produce transgenic plants capable of synthesizing natural rubber, including tobacco and sunflower. These efforts are focused on sequencing the subunits of the rubber transferase enzyme complex in order to transfect these genes into other plants.

## Membrane-associated Transferases

Many transferases associate with biological membranes as peripheral membrane proteins or anchored to membranes through a single transmembrane helix, for example numerous glycosyltransferases in Golgi apparatus. Some others are multi-span transmembrane proteins, for example certain Oligosaccharyltransferases or microsomal glutathione S-transferase from MAPEG family.

## Hydrolases

Hydrolase is a class of hydrolytic enzymes that are commonly used as biochemical catalysts utilizing water to break a chemical bond in order to divide a large molecule into two smaller ones. Hydrolases are pivotal for the body since they digest large molecules into fragments for synthesis, excrete waste materials, and provide carbon sources for the production of energy, during which many biopolymers are converted to monomers. Some hydrolases could release energy when they take effect. A number of hydrolases, especially proteases, are connected with biological membranes as peripheral membrane proteins or affixed through a single transmembrane helix. Some others are multi-span transmembrane ones. The names of hydrolases are systematically formed as "substrate hydrolase, while the commonly adopted names are typically in a form of "substratease."

In biochemistry, a hydrolase catalyzes the hydrolysis of a chemical bond like the following reaction:

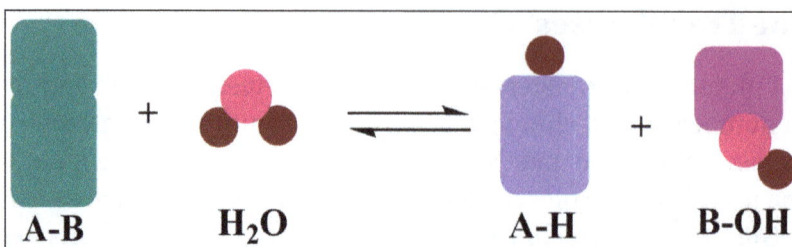

$$A\text{-}B \quad + \quad H_2O \quad \rightleftharpoons \quad A\text{-}H \quad + \quad B\text{-}OH$$

Hydrolysis reaction catalyzed by hydrolase.

# Classification

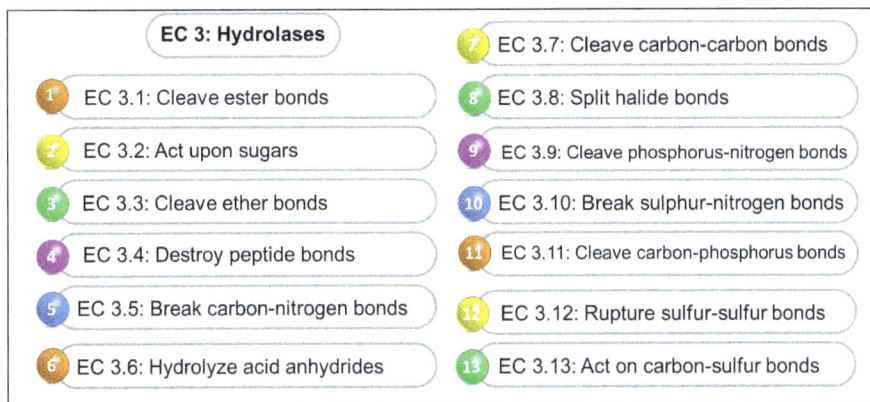

| EC 3: Hydrolases | EC 3.7: Cleave carbon-carbon bonds |
|---|---|
| 1. EC 3.1: Cleave ester bonds | 8. EC 3.8: Split halide bonds |
| 2. EC 3.2: Act upon sugars | 9. EC 3.9: Cleave phosphorus-nitrogen bonds |
| 3. EC 3.3: Cleave ether bonds | 10. EC 3.10: Break sulphur-nitrogen bonds |
| 4. EC 3.4: Destroy peptide bonds | 11. EC 3.11: Cleave carbon-phosphorus bonds |
| 5. EC 3.5: Break carbon-nitrogen bonds | 12. EC 3.12: Rupture sulfur-sulfur bonds |
| 6. EC 3.6: Hydrolyze acid anhydrides | 13. EC 3.13: Act on carbon-sulfur bonds |

Hydrolases belong to EC 3 in the EC classification system and can be further grouped into thirteen subclasses on the basis of the bonds they act upon. EC 3.1 represents a kind of enzymes rupturing ester bonds, which are called esterases. Some common esterases include nucleosidases, phosphatases, proteases, and lipases, among which phosphatases cut phosphate groups off molecules. Acetylcholine esterase is a potent neurotransmitter for voluntary muscle and it as one of the most crucial esterases contributes to the transform of the neuron impulse into acetic acid after it degrades acetylcholine into choline and acetic acid. Some dangerous toxins such as the exotoxin and saxitoxin could impede with the action of cholinesterase, and many nerve agents react by hindering the hydrolytic efficacy of cholinesterase. Nucleosidases are capable of hydrolyzing the bonds of nucleotides. Glycerides could be hydrolyzed by lipases, which also make contribution to the breakdown of fats, lipoproteins and other larger molecules into smaller molecules like fatty acids that are used for synthesis and as a source of energy. Hydrolases in EC 3.2 mainly act upon sugars such as DNA glycosylases and glycoside hydrolase. Acetic acid has become a nice intermediate for glycolysis catalyzed by glycosidases that chop sugar molecules into carbohydrates and peptidases hydrolyze peptide bonds. EC 3.3 includes ether bonds destroying enzymes. EC 3.4 covers hydrolases that act upon peptide bonds like proteases and peptidases. For example, acylpeptide hydrolase as a member of the peptidase family could deacetylate the acetylated N-terminus of polypeptides. Some other type of hydrolases comprise enzymes breaking carbon-nitrogen bonds, not peptide bonds, acid anhydrides (acid anhydride hydrolases, including helicases and GTPase), carbon-carbon bonds, halide bonds, phosphorus-nitrogen bonds, sulphur-nitrogen bonds, carbon-phosphorus bonds, sulfur-sulfur bonds, and carbon-sulfur bonds, with EC number sequentially ranging from 3.5 to 3.13.

## Various Functions of Various Hydrolases

Hydrolases could participate in a variety of biological procedures due to their diversification of acting position. Hydrolase expressed by Lactobacillus jensenii in the human gut could stimulate liver to secrete bile salts which facilitate the digestion of food. It has been suggested that the activity of bile salt hydrolase commonly found among intestinal microbiota could increase hydrogel-forming potencies of certain bile salts, whose occurrence in physiological conditions of human gut is thought to be able to enhance bacterial potnetiality to colonize the gastrointestinal tract and their survival rate in this specific ecological niche. Furthermore, new information about the activity of bile salt hydrolase in bacteria can be beneficial to more consciously make use of living microorganisms as food additives and also serves as guidance in the development of new medicines for the prevention and treatment of gastrointestinal disorders. Acylpeptide hydrolase found in human erythrocytes could be potentially treated as a biomarker for low dose exposure to organophosphorus in humans. Glycoside hydrolases play unique roles in various biological processes like the metabolism of cell wall, the biosynthesis of glycans, signalling, plant defence, and the mobilization of storage reserves. It has been found that a single residue in plant GH32, equivalent to Asp239 in AtcwINV1, seems to be important for sucrose stabilization in the active site and indispensable in determining the specificity of sucrose donor. The leukotriene A4 ($LTA_4$) hydrolase could act upon epoxide in the final step in the biosynthesis of leukotriene $B_4$ that occupies a position in a variety of acute and chronic inflammatory diseases. $LTA_4$ hydrolase as a bifunctional zinc metalloenzyme is a significant enzyme in the 5-lioxygenase pathway and possesses a peptide-cleaving activity. Nucleoside hydrolase plays a central role in the purine salvage pathway and also functions as a primary target for the explorations of anti-parasitic drugs.

## Lyases

Lyase, is an enzyme catalytically aiding in breaking various chemical bonds by means of an "elimination" reaction, other than hydrolysis and oxidation. This reaction often results in the formation of a new cyclic structure or a new double bond, and a reverse reaction called a "Michael addition" might also possibly happen under the catalysis of lyase. To obtain either a double bond or a new ring, lyase acts upon the single substrate and a molecule is eliminated. Lyases are different from other enzymes for only one substrate is required for the reaction in one direction, but two substrates are essential for the reverse reaction. Lyases can be commonly observed in the reactions of the Citric Acid Cycle (Krebs cycle) and in glycolysis. In glycolysis, aldolase could readily and reversibly degrade fructose 1,6-bisphosphate into the products glyceraldehyde 3-phosphate and dihydroxyacetone phosphate, which is an example of a lyase cleaving carbon-carbon bonds. Lyase works without the necessary requirements for cofactor recycling and gives an absolute stereospecificity with a theoretical yield of 100%, being much more efficient compared with enantiomeric resolutions of only 50% productive rate. Therefore, considerable researches have been addicted to the exploration of lyases as biocatalysts to synthesize optically active compounds, which have also been already found application in a few large commercial processes. Lyases are systematically named as "substrate group-lyase", such as decarboxylase, dehydratase, aldolase, etc.

## Classification

In the EC number classification of enzymes, EC 4 could represent lyases, which can be further classified into seven subclasses. Lyases in EC 4.1 cleave carbon-carbon bonds, and include decarboxylases (EC 4.1.1), aldehyde lyases (EC 4.1.2) facilitating the reverse reaction of aldol condensations, oxo acid lyases (EC 4.1.3) that catalyzes the cleavage of many 3-hydroxy acids, and others (EC 4.1.99). EC 4.2 contains a group of lyases that break carbon-oxygen bonds, such as dehydratases. Hydro-lyases being a part of carbon-oxygen lyases could facilitate the cleavage of C-O bonds by the elimination of water. Some other carbon-oxygen lyases promote the elimination of a phosphate or the removal of an alcohol from a polysaccharide. Lyases cleaving carbon-nitrogen bonds are sorted into EC 4.3. They could release ammonia with powerful cleaving ability and simultaneously form a double bond or ring. Some of these enzymes can also help to eliminate an amine or amide group. EC 4.4 represents lyases that split carbon-sulfur bonds, which could eliminate or substitute dihydrogen sulfide ($H_2S$) from a reaction. Carbon-halide bonds cleaving enzymes are lyases in EC 4.5 and that utilize an action mode that removes hydrochloric acid from a synthetic pesticide dichloro-diphenyl-trichloroethane (DDT). EC 4.6 comprises lyases fracturing phosphorus-oxygen bonds, like adenylyl cyclase and guanylyl cyclase, and they eliminate diphosphate from nucleotide triphosphates. EC 4.99 is a group of other lyases.

## Substrate Specificity

Narrow substrate specificity is usually considered to be a drawback for the commercialization of an enzyme in that it greatly restricts the flexibility of an enzyme as an assistant in the production of related compounds. Lyases are generally, but not always, found with narrow substrate specificity. Most hydratases and ammonia-lyases indeed possess quite narrow substrate specificity, while the substrate specificity for aldolases, decarboxylases and oxynitrilases is much broader. It is noteworthy here that the substrate specificity of a specific lyase varies depending on its source. However, it is not an absolute prerequisite for enzymes to own unrestricted substrate specificity for their commercial exploitation. In fact, there are several of the lyases in commercial use bearing a rather narrow substrate spectrum.

## Cofactor Requirements

The commercial potential of enzymes can be severely limited by the requirement for expensive cofactors. Since the addition catalyzed by lyase does not implicate a mere oxidation or reduction, it is not an essential requirement for cofactors. However, up to now, most of the lyases identified do require cofactors, which are involved in stabilization of reaction intermediates, polarization of the substrate, substrate binding, temporary binding of the nucleophile, and so on. The majority of these cofactors are not very expensive, and covalently bound to the enzyme. Thereby, the cofactors of lyases do not establish a barrier to their commercialization. The requirements for cofactors of lyases are varied according to their different sources.

## Isomerases

Isomerases are a general class of enzymes that convert a molecule from one isomer to another. Isomerases facilitate intramolecular rearrangements in which bonds are broken and formed. The general form of such a reaction is as follows:

A–B → B–A

There is only one substrate yielding one product. This product has the same molecular formula as the substrate but differs in bond connectivity or spatial arrangement. Isomerases catalyze reactions across many biological processes, such as in glycolysis and carbohydrate metabolism.

## Isomerization

Examples of isomers

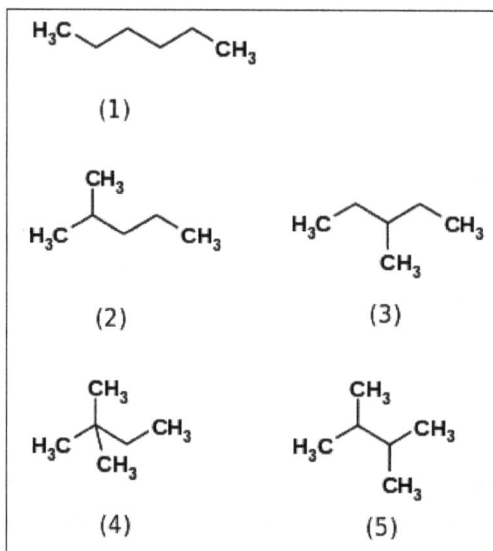

The structural isomers of hexane

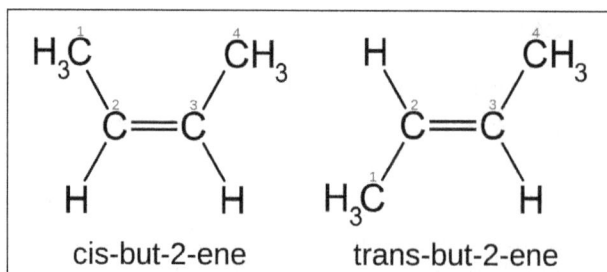

Cis-2-butene and Trans-2-butene

$$
\begin{array}{ccc}
\text{CHO} & \Big| & \text{CHO} \\
| & & | \\
\text{H—C—OH} & & \text{HO—C—H} \\
| & & | \\
\text{HO—C—H} & & \text{HO—C—H} \\
| & & | \\
\text{H—C—OH} & & \text{H—C—OH} \\
| & & | \\
\text{H—C—OH} & & \text{H—C—OH} \\
| & & | \\
\text{CH}_2\text{OH} & & \text{CH}_2\text{OH} \\
\text{D-Glucose} & & \text{D-Mannose}
\end{array}
$$

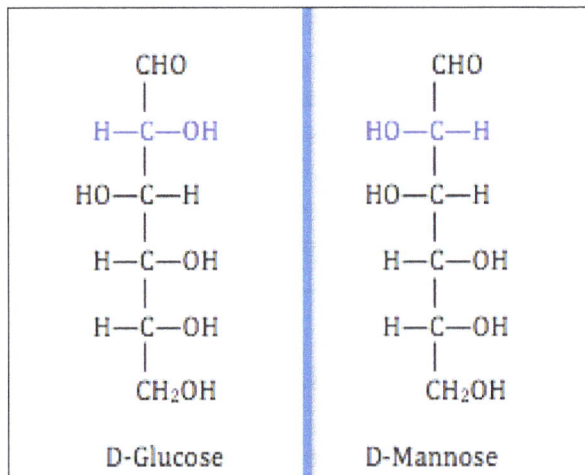

Epimers: D-glucose and D-mannose

Isomerases catalyze changes within one molecule. They convert one isomer to another, meaning that the end product has the same molecular formula but a different physical structure. Isomers themselves exist in many varieties but can generally be classified as structural isomers or stereoisomers. Structural isomers have a different ordering of bonds and/or different bond connectivity from one another, as in the case of hexane and its four other isomeric forms (2-methylpentane, 3-methylpentane, 2,2-dimethyl-butane, and 2,3-dimethylbutane).

Stereoisomers have the same ordering of individual bonds and the same connectivity but the three-dimensional arrangement of bonded atoms differ. For example, 2-butene exists in two isomeric forms: *cis*-2-butene and *trans*-2-butene. The sub-categories of isomerases containing racemases, epimerases and cis-trans isomers are examples of enzymes catalyzing the interconversion of stereoisomers. Intramolecular lyases, oxidoreductases and transferases catalyze the interconversion of structural isomers.

The prevalence of each isomer in nature depends in part on the isomerization energy, the difference in energy between isomers. Isomers close in energy can interconvert easily and are often seen in comparable proportions. The isomerization energy, for example, for converting from a stable *cis* isomer to the less stable *trans* isomer is greater than for the reverse reaction, explaining why in the absence of isomerases or an outside energy source such as ultraviolet radiation a given *cis* isomer tends to be present in greater amounts than the *trans* isomer. Isomerases can increase the reaction rate by lowering the isomerization energy.

Calculating isomerase kinetics from experimental data can be more difficult than for other enzymes because the use of product inhibition experiments is impractical. That is, isomerization is not an irreversible reaction since a reaction vessel will contain one substrate and one product so the typical simplified model for calculating reaction kinetics does not hold. There are also practical difficulties in determining the rate-determining step at high concentrations in a single isomerization. Instead, tracer perturbation can overcome these technical difficulties if there are two forms of the unbound enzyme. This technique uses isotope exchange to measure indirectly the interconversion of the free enzyme between its two forms. The radiolabeled substrate and product diffuse in a time-dependent manner. When the system reaches equilibrium the addition of unlabeled substrate perturbs or unbalances it. As equilibrium is established again, the radiolabeled substrate and product are tracked to determine energetic information.

The earliest use of this technique elucidated the kinetics and mechanism underlying the action of phosphoglucomutase, favoring the model of indirect transfer of phosphate with one intermediate and the direct transfer of glucose. This technique was then adopted to study the profile of proline racemase and its two states: the form which isomerizes L-proline and the other for D-proline. At high concentrations it was shown that the transition state in this interconversion is rate-limiting and that these enzyme forms may differ just in the protonation at the acidic and basic groups of the active site.

## Nomenclature

Generally, "the names of isomerases are formed as "*substrate* isomerase" (for example, enoyl CoA isomerase), or as "*substrate type of isomerase*" (for example, phosphoglucomutase)."

## Classification

Enzyme-catalyzed reactions each have a uniquely assigned classification number. Isomerase-catalyzed reactions have their own EC category: EC 5. Isomerases are further classified into six subclasses:

## Racemases and Epimerases

This category (EC 5.1) includes (racemases) and epimerases). These isomerases invert stereochemistry at the target chiral carbon. Racemases act upon molecules with one chiral carbon for inversion of stereochemistry, whereas epimerases target molecules with multiple chiral carbons and act upon one of them. A molecule with only one chiral carbon has two enantiomeric forms, such as serine having the isoforms D-serine and L-serine differing only in the absolute configuration about the chiral carbon. A molecule with multiple chiral carbons has two forms at each chiral carbon. Isomerization at one chiral carbon of several yields epimers, which differ from one another in absolute configuration at just one chiral carbon. For example, D-glucose and D-mannose differ in configuration at just one chiral carbon. This class is further broken down by the group the enzyme acts upon.

| Racemases and epimerases: | | |
|---|---|---|
| EC number | Description | Examples |
| EC 5.1.1 | Acting on Amino Acids and Derivative | alanine racemase, methionine racemase |
| EC 5.1.2 | Acting on Hydroxy Acids and Derivatives | lactate racemase, tartrate epimerase |
| EC 5.1.3 | Acting on Carbohydrates and Derivatives | ribulose-phosphate 3-epimerase, UDP-glucose 4-epimerase |
| EC 5.1.99 | Acting on Other Compounds | methylmalonyl CoA epimerase, hydantoin racemase |

## Cis-trans Isomerases

This category (EC 5.2) includes enzymes that catalyze the isomerization of cis-trans isomers. Alkenes and cycloalkanes may have cis-trans stereoisomers. These isomers are not distinguished by absolute configuration but rather by the position of substituent groups relative to a plane of reference, as across a double bond or relative to a ring structure. *Cis* isomers have substituent groups on the same side and *trans* isomers have groups on opposite sides.

This category is not broken down any further. All entries presently include:

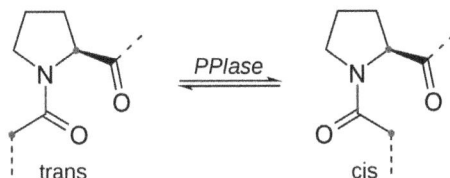

Conversion mediated by peptidylprolyl isomerase (PPIase).

| Cis-trans isomerases: | |
|---|---|
| EC number | Examples |
| EC 5.2.1.1 | Maleate isomerase |
| EC 5.2.1.2 | Maleylacetoacetate isomerase |
| EC 5.2.1.4 | Maleylpyruvate isomerase |
| EC 5.2.1.5 | Linoleate isomerase |
| EC 5.2.1.6 | Furylfuramide isomerase |
| EC 5.2.1.8 | Peptidylprolyl isomerase |
| EC 5.2.1.9 | Farnesol 2-isomerase |
| EC 5.2.1.10 | 2-chloro-4-carboxymethylenebut-2-en-1,4-olide isomerase |
| EC 5.2.1.12 | Zeta-carotene isomerase |
| EC 5.2.1.13 | Prolycopene isomerase |
| EC 5.2.1.14 | Beta-carotene isomerase |

## Intramolecular Oxidoreductases

This category (EC 5.3) includes intramolecular oxidoreductases. These isomerases cat-alyze the transfer of electrons from one part of the molecule to another. In other words, they catalyze the oxidation of one part of the molecule and the concurrent reduction of another part. Sub-categories of this class are:

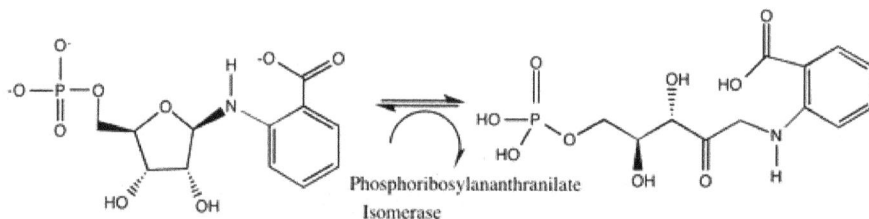

Reaction catalyzed by phosphoribosylanthranilate isomerase

| EC number | Description | Examples |
|---|---|---|
| EC 5.3.1 | Interconverting Aldoses and Ketoses | Triose-phosphate isomerase, Ribose-5-phosphate isomerase |
| EC 5.3.2 | Interconverting Keto- and Enol-Groups | Phenylpyruvate tautomerase, Oxaloacetate tautomerase |
| EC 5.3.3 | Transposing C=C Double Bonds | Steroid Delta-isomerase, L-dopachrome isomerase |
| EC 5.3.4 | Transposing S-S Bonds | Protein disulfide-isomerase |
| EC 5.3.99 | Other Intramolecular Oxidoreductases | Prostaglandin-D synthase, Allene-oxide cyclase |

## Intramolecular Transferases

This category (EC 5.4) includes intramolecular transferases (mutases). These isomer-ases catalyze the transfer of functional groups from one part of a molecule to another. Phosphotransferases (EC 5.4.2) were categorized as transferases (EC 2.7.5) with re-generation of donors until 1983. This sub-class can be broken down according to the functional group the enzyme transfers:

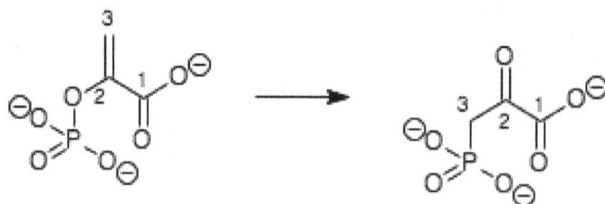

Reaction catalyzed by phosphoenolpyruvate mutase

| Intramolecular transferases: | | |
|---|---|---|
| **EC number** | **Description** | **Examples** |
| EC 5.4.1 | Transferring Acyl Groups | Lysolecithin acylmutase, Precorrin-8X methylmutase |
| EC 5.4.2 | Phosphotransferases (Phosphomutases) | Phosphoglucomutase, Phosphopentomutase |
| EC 5.4.3 | Transferring Amino Groups | Beta-lysine 5,6-aminomutase, Tyrosine 2,3-aminomutase |
| EC 5.4.4 | Transferring hydroxy groups | (hydroxyamino)benzene mutase, Isochorismate synthase |
| EC 5.4.99 | Transferring Other Groups | Methylaspartate mutase, Chorismate mutase |

## Intramolecular Lyases

This category (EC 5.5) includes intramolecular lyases. These enzymes catalyze "reactions in which a group can be regarded as eliminated from one part of a molecule, leaving a double bond, while remaining covalently attached to the molecule." Some of these catalyzed reactions involve the breaking of a ring structure.

This category is not broken down any further. All entries presently include:

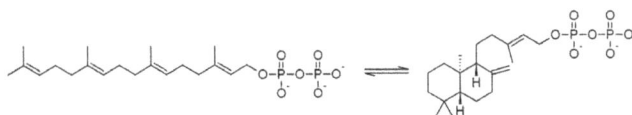

reaction catalyzed by ent-Copalyl diphosphate synthase

| Intramolecular lyases: | |
|---|---|
| **EC number** | **Examples** |
| EC 5.5.1.1 | Muconate cycloisomerase |
| EC 5.5.1.2 | 3-carboxy-cis,cis-muconate cycloisomerase |
| EC 5.5.1.3 | Tetrahydroxypteridine cycloisomerase |
| EC 5.5.1.4 | Inositol-3-phosphate synthase |
| EC 5.5.1.5 | Carboxy-cis,cis-muconate cyclase |
| EC 5.5.1.6 | Chalcone isomerase |
| EC 5.5.1.7 | Chloromuconate cycloisomerase |
| EC 5.5.1.8 | (+)-bornyl diphosphate synthase |
| EC 5.5.1.9 | Cycloeucalenol cycloisomerase |
| EC 5.5.1.10 | Alpha-pinene-oxide decyclase |
| EC 5.5.1.11 | Dichloromuconate cycloisomerase |
| EC 5.5.1.12 | Copalyl diphosphate synthase |
| EC 5.5.1.13 | Ent-copalyl diphosphate synthase |

| EC 5.5.1.14 | Syn-copalyl-diphosphate synthase |
| EC 5.5.1.15 | Terpentedienyl-diphosphate synthase |
| EC 5.5.1.16 | Halimadienyl-diphosphate synthase |
| EC 5.5.1.17 | (S)-beta-macrocarpene synthase |
| EC 5.5.1.18 | Lycopene epsilon-cyclase |
| EC 5.5.1.19 | Lycopene beta-cyclase |
| EC 5.5.1.20 | Prosolanapyrone-III cycloisomerase |
| EC 5.5.1.n1 | D-ribose pyranase |

## Mechanisms of Isomerases

## Ring Expansion and Contraction via Tautomers

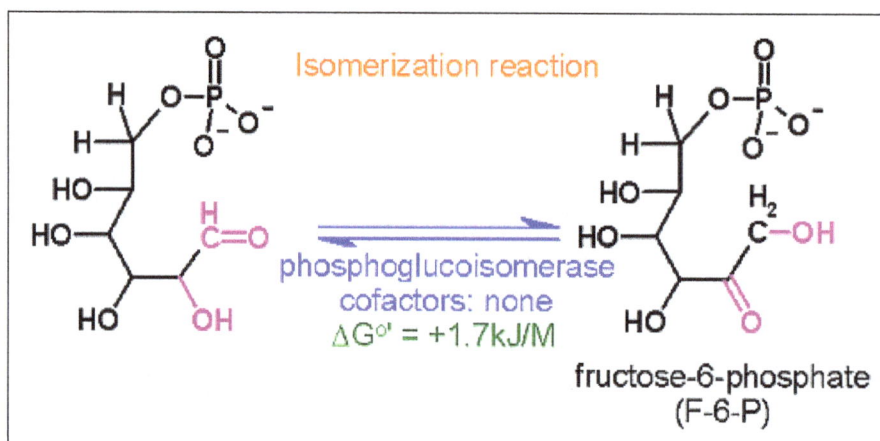

The isomerization of glucose-6-phosphate by glucose-6-phosphate isomerase

A classic example of ring opening and contraction is the isomerization of glucose (an aldehyde with a six-membered ring) to fructose (a ketone with a five-membered ring). The conversion of D-glucose-6-phosphate to D-fructose-6-phosphate is catalyzed by glucose-6-phosphate isomerase, an intramolecular oxidoreductase. The overall reaction involves the opening of the ring to form an aldose via acid/base catalysis and the subsequent formation of a cis-endiol intermediate. A ketose is then formed and the ring is closed again.

Glucose-6-phosphate first binds to the active site of the isomerase. The isomerase opens the ring: its His388 residue protonates the oxygen on the glucose ring (and thereby breaking the $O5-C1$ bond) in conjunction with Lys518 deprotonating the C1 hydroxyl oxygen. The ring opens to form a straight-chain aldose with an acidic C2 proton. The C3-C4 bond rotates and Glu357 (assisted by His388) depronates C2 to form a double bond between C1 and C2. A cis-endiol intermediate is created and the C1 oxygen is protonated by the catalytic residue, accompanied by the deprotonation of the endiol C2 oxygen. The straight-chain ketose is formed. To close the fructose ring, the reverse of ring opening occurs and the ketose is protonated.

# Epimerization

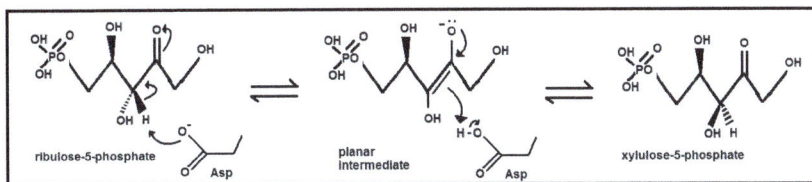

The conversion of ribulose-5-phosphate to xylulose-5-phosphate

An example of epimerization is found in the Calvin cycle when D-ribulose-5-phosphate is converted into D-xylulose-5-phosphate by ribulose-phosphate 3-epimerase. The substrate and product differ only in stereochemistry at the third carbon in the chain. The underlying mechanism involves the deprotonation of that third carbon to form a reactive enolate intermediate. The enzyme's active site contains two Asp residues. After the substrate binds to the enzyme, the first Asp deprotonates the third carbon from one side of the molecule. This leaves a planar sp²-hybridized intermediate. The second Asp is located on the opposite side of the active side and it protonates the molecule, effectively adding a proton from the back side. These coupled steps invert stereochemistry at the third carbon.

# Intramolecular Transfer

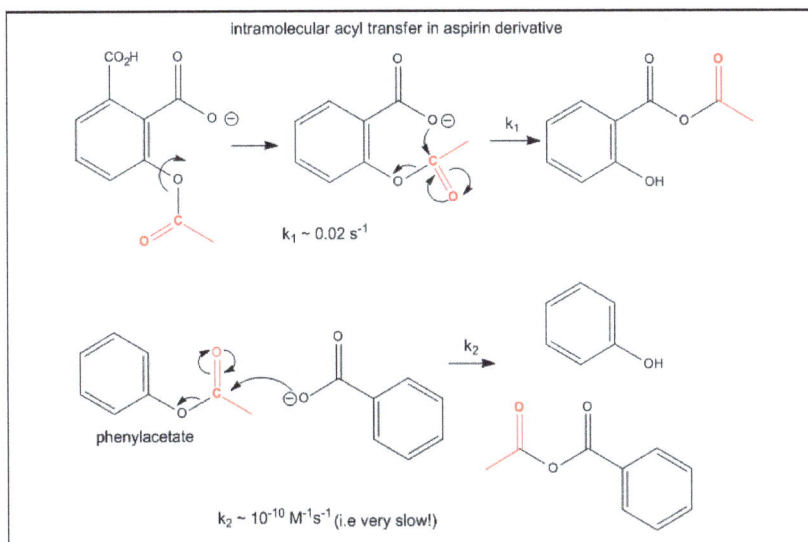

Chorismate mutase is an intramolecular transferase and it catalyzes the conversion of chorismate to prephenate, used as a precursor for L-tyrosine and L-phenylalanine in some plants and bacteria. This reaction is a Claisen rearrangement that can proceed with or without the isomerase, though the rate increases $10^6$ fold in the presence of chorismate mutase. The reaction goes through a chair transition state with the substrate in a trans-diaxial position. Experimental evidence indicates that the isomerase selectively binds the chair transition state, though the exact mechanism of catalysis is not known. It is thought that this binding stabilizes the transition state through electrostatic effects, accounting for the dramatic increase in the reaction rate in the presence of the mutase or upon addition of a specifically-placed cation in the active site.

## Intramolecular Oxidoreduction

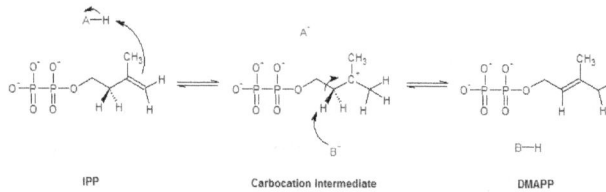

IPP                      Carbocation Intermediate                      DMAPP

Conversion by IPP isomerase

Isopentenyl-diphosphate delta isomerase type I (also known as IPP isomerase) is seen in cholesterol synthesis and in particular it catalyzes the conversion of isopentenyl diphosphate (IPP) to dimethylallyl diphosphate (DMAPP). In this isomerization reaction a stable carbon-carbon double bond is rearranged top create a highly electrophilic allylic isomer. IPP isomerase catalyzes this reaction by the stereoselective antarafacial transposition of a single proton. The double bond is protonated at C4 to form a tertiary carbocation intermediate at C3. The adjacent carbon, C2, is deprotonated from the opposite face to yield a double bond. In effect, the double bond is shifted over.

## Role of Isomerase in Human Disease

Isomerase plays a role in human disease. Deficiencies of this enzyme can cause disorders in humans.

## Phosphohexose Isomerase Deficiency

Phosphohexose Isomerase Dificiency (PHI) is also known as phosphoglucose isomerase deficiency or Glucose-6-phosphate isomerase deficiency, and is a hereditary enzyme deficiency. PHI is the second most frequent erthoenzyopathy in glycolysis besides pyruvate kinase deficiency, and is associated with non-spherocytic haemolytic anaemia of variable severity. This disease is centered on the glucose-6-phosphate protein. This protein can be found in the secretion of some cancer cells. PHI is the result of a dimeric enzyme that catalyses the reversible interconversion of fructose-6-phosphate and gluose-6-phosphate.

PHI is a very rare disease with only 50 cases reported in literature to date.

Diagnosis is made on the basis of the clinical picture in association with biochemical studies revealing erythrocyte GPI deficiency (between 7 and 60% of normal) and identification of a mutation in the GPI gene by molecular analysis.

The deficiency of phosphohexose isomerase can lead to a condition referred to as hemolytic syndrome. As in humans, the hemolytic syndrome, which is characterized by a diminished erythrocyte number, lower hematocrit, lower hemoglobin, higher number of reticulocytes and plasma bilirubin concentration, as well as increased liver- and spleen-somatic indices, was exclusively manifested in homozygous mutants.

## Triosephosphate Isomerase Deficiency

The disease referred to as triosephosphate isomerase deficiency (TPI), is a severe autosomal recessive inherited multisystem disorder of glycolyic metabolism. It is characterized by hemolytic anemia and neurodegeneration, and is caused by anaerobic metabolic dysfunction. This dysfunction results from a missense mutation that effects the encoded TPI protein. The most common mutation is the substitution of gene, Glu104Asp, which produces the most severe phenotype, and is responsible for approximately 80% of clinical TPI deficiency.

TPI deficiency is very rare with less than 50 cases reported in literature. Being an autosomal recessive inherited disease, TPI deficiency has a 25% recurrence risk in the case of heterozygous parents. It is a congenital disease that most often occurs with hemolytic anemia and manifests with jaundice. Most patients with TPI for Glu104Asp mutation or heterozygous for a TPI null allele and Glu104Asp have a life expectancy of infancy to early childhood. TPI patients with other mutations generally show longer life expectancy. To date, there are only two cases of individuals with TPI living beyond the age of 6. These cases involve two brothers from Hungary, one who did not develop neurological symptoms until the age of 12, and the older brother who has no neurological symptoms and suffers from anemia only.

Individuals with TPI show obvious symptoms after 6–24 months of age. These symptoms include: dystonia, tremor, dyskinesia, pyramidal tract signs, cardiomyopathy and spinal motor neuron involvement. Patients also show frequent respiratory system bacterial infections.

TPI is detected through deficiency of enzymatic activity and the build-up of dihyroxyacetone phosphate(DHAP), which is a toxic substrate, in erythrocytes. This can be detected through physical examination and a series of lab work. In detection, there is generally myopathic changes seen in muscles and chronic axonal neuropathy found in the nerves. Diagnosis of TPI can be confirmed through molecular genetics. Chorionic villus DNA analysis or analysis of fetal red cells can be used to detect TPI in antenatal diagnosis.

Treatment for TPI is not specific, but varies according to different cases. Because of the range of symptoms TPI causes, a team of specialist may be needed to provide treatment to a single individual. That team of specialists would consists of pediatricians, cardiologists, neurologists, and other healthcare professionals, that can develop a comprehensive plan of action.

Supportive measures such as red cell transfusions in cases of severe anaemia can be taken to treat TPI as well. In some cases, spleen removal (splenectomy) may improve the anaemia. There is no treatment to prevent progressive neurological impairment of any other non-haematological clinical manifestation of the diseases.

## Industrial Applications

By far the most common use of isomerases in industrial applications is in sugar manufacturing. Glucose isomerase (also known as xylose isomerase) catalyzes the conversion of D-xylose and D-glucose to D-xylulose and D-fructose. Like most sugar isomerases, glucose isomerase catalyzes the interconversion of aldoses and ketoses.

The conversion of glucose to fructose is a key component of high-fructose corn syrup production. Isomerization is more specific than older chemical methods of fructose production, resulting in a higher yield of fructose and no side products. The fructose produced from this isomerization reaction is purer with no residual flavors from contaminants. High-fructose corn syrup is preferred by many confectionery and soda manufacturers because of the high sweetening power of fructose (twice that of sucrose), its relatively low cost and its inability to crystallize. Fructose is also used as a sweetener for use by diabetics. Major issues of the use of glucose isomerase involve its inactivation at higher temperatures and the requirement for a high pH (between 7.0 and 9.0) in the reaction environment. Moderately high temperatures, above 70 °C, increase the yield of fructose by at least half in the isomerization step. The enzyme requires a divalent cation such as $Co^{2+}$ and $Mg^{2+}$ for peak activity, an additional cost to manufacturers. Glucose isomerase also has a much higher affinity for xylose than for glucose, necessitating a carefully controlled environment.

The isomerization of xylose to xylulose has its own commercial applications as interest in biofuels has increased. This reaction is often seen naturally in bacteria that feed on decaying plant matter. Its most common industrial use is in the production of ethanol, achieved by the fermentation of xylulose. The use of hemicellulose as source material is very common. Hemicellulose contains xylan, which itself is composed of xylose in β(1,4) linkages. The use of glucose isomerase very efficiently converts xylose to xylulose, which can then be acted upon by fermenting yeast. Overall, extensive research in genetic engineering has been invested into optimizing glucose isomerase and facilitating its recovery from industrial applications for re-use.

Glucose isomerase is able to catalyze the isomerization of a range of other sugars, including D-ribose, D-allose and L-arabinose. The most efficient substrates are those similar to glucose and xylose, having equatorial hydroxyl groups at the third and fourth carbons. The current model for the mechanism of glucose isomerase is that of a hydride shift based on X-ray crystallography and isotope exchange studies.

## Membrane-associated Isomerases

Some isomerases associate with biological membranes as peripheral membrane proteins or anchored through a single transmembrane helix, for example isomerases with the thioredoxin domain, and certain prolyl isomerases.

# Hexokinase

A hexokinase is an enzyme that phosphorylates a six-carbon sugar, a hexose, to a hexose phosphate. In most tissues and organisms, glucose is the most important substrate of hexokinases, and glucose-6-phosphate the most important product.

## Variation across Species

Genes that encode hexokinase have been discovered in each domain of life, ranging from bacteria, yeast, and plants to humans and other vertebrates. They are categorized as actin fold proteins, sharing a common ATP binding site core surrounded by more variable sequences that determine substrate affinities and other properties. Several hexokinase isoforms or isozymes providing different functions can occur in a single species. Hexokinase should not be confused with the liver's glucokinase. While hexokinase is capable of phosphorylating several hexoses, glucokinase acts with a 50-fold lower substrate affinity, and its only substrate is glucose.

## Reaction

The intracellular reactions mediated by hexokinases can be typified as:

$$\text{Hexose-CH}_2\text{OH} + \text{MgATP}^{2-} \rightarrow \text{Hexose-CH}_2\text{O-PO}_3^{2-} + \text{MgADP-} + \text{H}^+$$

where hexose-$CH_2OH$ represents any of several hexoses (like glucose) that contain an accessible -$CH_2OH$ moiety.

## Consequences of Hexose Phosphorylation

Phosphorylation of a hexose such as glucose often limits it to a number of intracellular metabolic processes, such as glycolysis or glycogen synthesis. Phosphorylation makes hexose unable to move or be transported out of the cell.

## Size of Different Isoforms

Most bacterial hexokinases are approximately 50 kD in size. Multicellular organisms such as plants and animals often have more than one hexokinase isoform. Most are about 100 kD in size and consist of two halves (N and C terminal), which share much sequence homology. This suggests an evolutionary origin by duplication and fusion of a 50kD ancestral hexokinase similar to those of bacteria.

## Types of Mammalian Hexokinase

There are four important mammalian hexokinase isozymes (EC 2.7.1.1) that vary in subcellular locations and kinetics with respect to different substrates and conditions, and physiological function. They are designated hexokinases I, II, III, and IV or hexokinases A, B, C, and D.

## Hexokinases I, II and III

Hexokinases I, II, and III are referred to as "low-$K_m$" isozymes because of a high affinity for glucose even at low concentrations (below 1 mM). Hexokinases I and II follow Michaelis-Menten kinetics at physiologic concentrations of substrates. All three are strongly inhibited by their product, glucose-6-phosphate. Molecular weights are around 100 kD. Each consists of two similar 50kD halves, but only in hexokinase II do both halves have functional active sites.

- Hexokinase I/A is found in all mammalian tissues, and is considered a "housekeeping enzyme," unaffected by most physiological, hormonal, and metabolic changes.

- Hexokinase II/B constitutes the principal regulated isoform in many cell types and is increased in many cancers.

- Hexokinase III/C is substrate-inhibited by glucose at physiologic concentrations. Little is known about the regulatory characteristics of this isoform.

## Hexokinase IV Glucokinase

Mammalian hexokinase IV, also referred to as glucokinase, differs from other hexokinases in kinetics and functions.

- The location of the phosphorylation on a subcellular level occurs when glucokinase translocates between the cytoplasm and nucleus of liver cells. Glucokinase can only phosphorylate glucose if the concentration of this substrate is high enough; its Km for glucose is 100 times higher than that of hexokinases I, II, and III.

- It is monomeric, about 50kD, displays positive cooperativity with glucose, and is not allosterically inhibited by its product, glucose-6-phosphate.

- It is present in the liver, pancreas, hypothalamus, small intestine, and perhaps certain other neuroendocrine cells, and plays an important regulatory role in carbohydrate metabolism.

- In the beta cells of the pancreatic islets, it serves as a glucose sensor to control insulin release, and similarly controls glucagon release in the alpha cells.

- In hepatocytes of the liver, glucokinase responds to changes of ambient glucose levels by increasing or reducing glycogen synthesis.

## Hexokinase in Glycolysis

Glucose is unique in that it can be used as an energy source by all cells in both the presence and absence of molecular oxygen ($O_2$). Glucose metabolism via the metabolic pathway known as glycolysis is importantly coupled to mitochondrial oxidative metabolism in the presence of oxygen, but glycolysis can also generate ATP in its absence. The first step in this sequence of reactions is the phosphorylation of glucose by hexokinase to prepare it for later breakdown in order to provide energy.

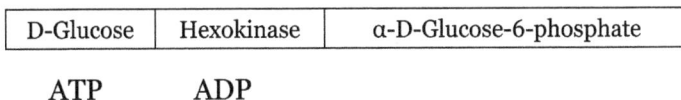

| D-Glucose | Hexokinase | α-D-Glucose-6-phosphate |
|-----------|------------|--------------------------|

ATP          ADP

By catalyzing the phosphorylation of glucose to yield glucose 6-phosphate - the first committed step of glucose metabolism - hexokinases importantly maintain the downhill concentration gradient permitting facilitated glucose transport into cells. This reaction also initiates all physiologically relevant pathways of glucose utilization, including glycolysis and the pentose phosphate pathway. The addition of a charged phosphate group at the 6-position of hexoses also ensures 'trapping' of glucose and 2-deoxyhexose glucose analogs (e.g. 2-deoxyglucose, and 2-fluoro-2-deoxyglucose) within cells, as charged hexose phosphates cannot easily cross the cell membrane.

## Association with Mitochondria

Hexokinases I and II can associate physically to the outer surface of the external membrane of mitochondria through specific binding to a porin, or voltage dependent anion channel. This association confers hexokinase direct access to ATP generated by mitochondria, which is one of the two substrates of hexokinase. Mitochondrial hexokinase is highly elevated in rapidly-growing malignant tumor cells, with levels up to 200 times higher than normal tissues. Mitochondrially-bound hexokinase has been demonstrated to be the driving force for the extremely high glycolytic rates that take place aerobically in tumor cells.

## Hydropathy Plot

The potential transmembrane portions of a protein can be detected by hydropathy analysis. A hydropathy analysis uses an algorithm that quantifies the hydrophobic character at each position along the polypeptide chain. One of the accepted hydropathy scales is that of Kyte and Doolittle which relies on the generation of hydropathy plots. In these plots, the negative numbers represent hydrophilic regions and the positive numbers represent hydrophobic regions on the y-axis. A potential transmembrane domain is about 20 amino acids long on the x-axis. Hexokinase has both hydrophilic and hydrophobic regions due to its folding. It appears as if hexokinase has one potential transmembrane domain located at about amino acid 400. Therefore, hexokinase is most likely not an integral membrane protein in yeast.

## Maltase

Maltese is also a carbohydrate-digesting enzyme that can be found naturally in sugars produced by the body when it breaks down starch.

- Additionally, it is a by-product when it comes to consuming sugar throughout several cooking processes, particularly during burning at high temperatures when the sugar changes colors from white to brown.

- Maltese is known to break down disaccharide maltose in 2 glucose molecules easily oxidized by the body in exchange for energy.

- Simply put, maltase is really important when it comes to the overall enzymatic process because it is used efficiently by the body to digest sugars and starch found under the shape of grains and other foods based on grains that we consume daily.

## Health Benefits

Maltase is known as an essential digestive enzyme found in people's mouths and saliva. It can ease digestion within the small intestine and the pancreas. The lack of maltase within the system might cause problems because the small intestine will have a more difficult job in breaking down starches and sugars. Thus, the enzyme can be of great help for the whole digestive system. It can help people benefit from a smooth bowel digestion.

Used within mucus membranes, the enzyme is included in the interior intestinal wall. Because it is located in the mouth, maltase works together with additional digestive carbohydrate enzymes to make sugars and starches simpler to digest. The process is halted and temporary reduced throughout more acidic digestion phases within the stomach; however, it is also resumed within the neutral pH of small intestines where maltase will be again secreted. The enzyme's vegetarian form is created through a natural process of fermentation known as Aspergillis oryzae.

- Maltase is an enzyme that can also stop and support chronic diarrhea. Several studies performed on patients suffering from diarrhea showed that enzyme deficiency triggered the condition. In addition, the lack of important enzymes within the body might lead to chronic diarrhea and the only remedy would be ingesting supplement rich in enzymes in order to reduce inflammation, infections and mucosal secretions in the gut.

- Maltase breaks down grain and it is shaped in a way where it can easily break the connection and set free the 2 glucoses pieces that are linked together. Hence, it can split maltose molecules extremely fast and efficiently.

- Maltase can work as a support and preventive mechanism for various digestive complaints in kids who suffer from autism. Advanced technology has managed to develop tremendously and thus, the use of enzymes like maltase could have beneficial effects. Several research studies have confirmed that numerous kids with autism showed a correlation in decreased intestinal disaccharide commotion. This is connected to the existence of digestive enzymes such as maltase within the gut.

## Trypsin

Trypsin is a proteolytic enzyme, important for the digestion of proteins. In humans, the protein is produced in its inactive form, trypsinogen, within the pancrease. Trypsinogen enters the small intestine, via the common bile duct, where it converted to active trypsin. Trypsin cleaves a terminal hexapeptide from trypsinogen to yield a single-chain [beta]-trypsin. Subsequent autolysis produces other active forms having two or more peptide chains. The two predominant forms of trypsin are [alpha]-trypsin, which has two peptide chains bound by disulfide bonds, and [beta]-trypsin.

Trypsin degrades proteins. As trypsin is itslf a protein, it is capable of digesting inself: a process called autolysis. Autolysis is important for the regulation of trypsin levels within living organisms. This regulation is assisted by $Ca^{2+}$ ions, which bind to trypsin (at the $Ca^{2+}$ binding loop) and protect the molecule against autolysis. In living organisms, autolysis is controlled and normally does not cause problems. However, when working with trypin in vitro, the process of autolysis often poses some problems. For in vitro processes that require the use of trypsin, such as working with cell cultures or manufacturing insulin, trypsin's degradation can become expensive as active trypsin gets "used up." Developing mutant trypsin that does not auto-degrade could be of great use for researchers.

There are several sites on the trypsin molecule at which autolysis is known to occur. Research has been done to investigate these sites, because the inability of trypsin to self-degrade has been linked to human hereditary pancreatitis. This deadly disease is believed to occur due to inappropriate activation of trypsin within the pancrease. This results in the autodigestion of pancreatic tissue. In this investigation, we will be studying rat (Rattus rattus) trypsin and a mutant form of rat trypsin, in which two autolytic sites have been removed. This mutant form of trypsin could help researchers understand hereditary pancreatitis and could be useful for research that is dependent upon significant use of active trypsin.

## Catalytic Triad

The enzymatic activity of trypsin is highly specific towards the positive side-chains of residues lysine (Lys) and arginine (Arg), cleaving a peptide at the carboxyl side of these residues, during a hydrolytic reaction. The catalytic triad of trypsin forms the active site of the enzyme. Three amino acid residues, His57, Asp102, and Cys195, are vital to the proteolytic function of the molecule. In the absence of a substrate protein, His57 is unprotonated. However, when the sulfur atom of Cys195 carries out a nucleophilic attack on the substrate, His57 accepts a proton from Cys195. The role of Asp102 is to stabilize the positively charged form of His57 in the transition state. When the substrate moves in and binds to Cys195, a tetrahedral transition state is formed, with the substrate's carbonyl oxygen becoming negatively charged, as it forms a single bond. In the next step, the amine component of the substrate forms a hydrogen bond to His57, while the acid component is covalently bound to Cys195. The amine component diffuses away, thus completing the acylation stage of the hydrolytic reaction. During the next step, a water molecule takes the place of the amine component of the original substrate. Then His57 draws a proton away from water, while the resulting OH$^-$ ion attacks the carbonyl carbon atom of the acyl group that is attached to Cys195 to form the next tetrahedral transition state. Finally, His57 donates the proton to the sulfur atom of Cys195, which then releases the acid component of the substrate.

## Mutant Trypsin

While rat (Rattus rattus) trypsin has thirteen potential trypsinsensitive sites (12 shown), there are two especially important autolytic sites that have been reported: Lys61-Ser62 and Arg117-Val118. It is known that autolysis of wild-type trypsin begins with the cleavage of the Arg117-Val118 peptide bond. In addition, the peptide segment between these two sites is part of the longest peptide chain not stabilized by disulfidebridges; this region may function as a built-in target for autolysis. As long as both ends of the peptide remain intact, the other cleavagesites within this region appear to be protected from hydrolysis.

Site-directed mutagenesis of Lys61 and Arg117 to Asn61 and Asn117, resulted in a trypsin mutant that was almost completely resistant to autolysis. In addition, these mutations did not significantly alter the catalytic efficiency of the enzyme.

The rate of autolysis is dependent upon $Ca^{2+}$ concentration; the mechanism by which $Ca^{2+}$ binding affects autolysis is still unknown. $Ca^{2+}$ has been shown to protect trypsin from autolysis. The $Ca^{2+}$-binding loop extends from Glu70 to Glu80, within the "self-destruction" segment. As the mutant trypsin was resistant to autolysis and does not need Ca2+ to protect it from autolysis, the protein was found to be almost completely insensitive to the presense of $Ca^{2+}$. Finally, the loss of activity of wild-type trypsin can be explaned by the disruption of this peptide segment, as two members of the catalytic triad, His57 and Asp102, are near or a part of this region.

This double mutant trypsin may prove useful in experiments in which autolysis of wild-type trypsin has caused problems. Finally, this mutant trypsin will be useful in understanding and treating human hereditary pancreatitis.

## Lactase

Lactase is an enzyme produced by many organisms. It is located in the brush border of the small intestine of humans and other mammals. Lactase is essential to the complete digestion of whole milk; it breaks down lactose, a sugar which gives milk its sweetness. Lacking lactase, a person consuming dairy products may experience the symptoms of lactose intolerance. Lactase can be purchased as a food supplement, and is added to milk to produce "lactose-free" milk products.

Lactase (also known as lactase-phlorizin hydrolase, or LPH), a part of the β-galactosidase family of enzymes, is a glycoside hydrolase involved in the hydrolysis of the disaccharide lactose into constituent galactose and glucose monomers. Lactase is present predominantly along the brush border membrane of the differentiated enterocytes lining the villi of the small intestine. In humans, lactase is encoded by the LCT gene.

## Uses

### Food Use

Lactase is an enzyme that some people are unable to produce in their small intestine. Without it they can't break down the natural lactose in milk, leaving them with diarrhea, gas and bloating when drinking regular milk. Technology to produce lactose-free milk, ice cream and yogurt was developed by the USDA Agricultural Research Service in 1985. This technology is used to add lactase to milk, thereby hydrolyzing the lactose naturally found in milk, leaving it slightly sweet but digestible by everyone. Without lactase, lactose intolerant people pass the lactose undigested to the colon where bacteria break it down, creating carbon dioxide and that leads to bloating and flatulence.

### Medical Use

Lactase supplements are sometimes used to treat lactose intolerance.

### Industrial Use

Lactase produced commercially can be extracted both from yeasts such as *Kluyveromyces fragilis* and *Kluyveromyces lactis* and from molds, such as *Aspergillus niger* and *Aspergillus oryzae*. Its primary commercial use, in supplements such as Lacteeze and Lactaid, is to break down lactose in milk to make it suitable for people with lactose intolerance, However, the U.S. Food and Drug Administration has not formally evaluated the effectiveness of these products.

Lactase is also used to screen for blue white colonies in the multiple cloning sites of various plasmid vectors in *Escherichia coli* or other bacteria.

## Mechanism

The optimum temperature for human lactase is about 37 °C for its activity and the optimum pH is 6.

In metabolism, the β-glycosidic bond in *D*-lactose is hydrolyzed to form *D*-galactose and *D*-glucose, which can be absorbed through the intestinal walls and into the bloodstream. The overall reaction that lactase catalyzes is $C_{12}H_{22}O_{11} + H_2O \rightarrow C_6H_{12}O_6 + C_6H_{12}O_6$ + heat.

The catalytic mechanism of *D*-lactose hydrolysis retains the substrate anomeric configuration in the products. While the details of the mechanism are uncertain, the stereochemical retention is achieved through a double displacement reaction. Studies of *E. coli* lactase have proposed that hydrolysis is initiated when a glutamate nucleophile on the enzyme attacks from the axial side of the galactosyl carbon in the β-glycosidic bond. The removal of the *D*-glucose leaving group may be facilitated by Mg-dependent acid catalysis. The enzyme is liberated from the α-galactosyl moiety upon equatorial nucleophilic attack by water, which produces *D*-galactose.

Substrate modification studies have demonstrated that the 3'-OH and 2'-OH moieties on the galactopyranose ring are essential for enzymatic recognition and hydrolysis. The 3'-hydroxy group is involved in initial binding to the substrate while the 2'- group is not necessary for recognition but needed in subsequent steps. This is demonstrated by the fact that a 2-deoxy analog is an effective competitive inhibitor ($K_i$ = 10mM). Elimination of specific hydroxyl groups on the glucopyranose moiety does not completely eliminate catalysis.

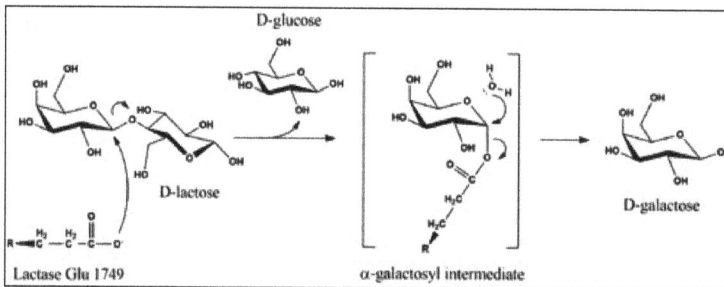

Lactase also catalyzes the conversion of phlorizin to phloretin and glucose.

## Structure and Biosynthesis

Preprolactase, the primary translation product, has a single polypeptide primary structure consisting of 1927 amino acids. It can be divided into five domains: (i) a 19-amino-acid cleaved signal sequence; (ii) a large prosequence domain that is not present in mature lactase; (iii) the mature lactase segment; (iv) a membrane-spanning hydrophobic anchor; and (v) a short hydrophilic carboxyl terminus. The signal sequence is cleaved in the endoplasmic reticulum, and the resulting 215-kDa pro-LPH is sent to the Golgi apparatus, where it is heavily glycosylated and proteolytically processed to its mature form. The prodomain has been shown to act as an intramolecular chaperone in the ER, preventing trypsin cleavage and allowing LPH to adopt the necessary 3-D structure to be transported to the Golgi apparatus.

Mature human lactase consists of a single 160-kDa polypeptide chain that localizes to the brush border membrane of intestinal epithelial cells. It is oriented with the N-terminus outside the cell and the C-terminus in the cytosol. LPH contains two catalytic glutamic acid sites. In the human enzyme, the lactase activity has been connected to Glu-1749, while Glu-1273 is the site of phlorizin hydrolase function.

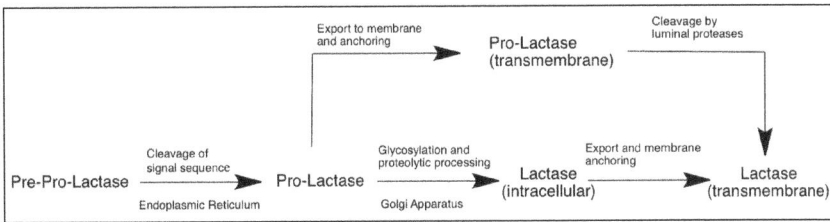

Schematic of processing and localization of human lactase translational product

## Genetic Expression and Regulation

Lactase is encoded by a single genetic locus on chromosome 2. It is expressed exclusively by mammalian small intestine enterocytes and in very low levels in the colon during fetal development. Humans are born with high levels of lactase expression. In most of the world's population, lactase transcription is down-regulated after weaning, resulting in diminished lactase expression in the small intestine, which causes the common symptoms of adult-type hypolactasia, or lactose intolerance.

Some population segments exhibit lactase persistence resulting from a mutation that is postulated to have occurred 5,000–10,000 years ago, coinciding with the rise of cattle domestication. This mutation has allowed almost half of the world's population to metabolize lactose without symptoms. Studies have linked the occurrence of lactase persistence to two different single-nucleotide polymorphisms about 14 and 22 kilobases upstream of the 5'-end of the LPH gene. Both mutations, C→T at position -13910 and G→A at position -22018, have been independently linked to lactase persistence.

The lactase promoter is 150 base pairs long and is located just upstream of the site of transcription initiation. The sequence is highly conserved in mammals, suggesting that critical cis-transcriptional regulators are located nearby. Cdx-2, HNF-1α, and GATA have been identified as transcription factors. Studies of hypolactasia onset have demonstrated that despite polymorphisms, little difference exists in lactase expression in infants, showing that the mutations become increasingly relevant during development. Developmentally regulated DNA-binding proteins may down-regulate transcription or destabilize mRNA transcripts, causing decreased LPH expression after weaning.

## DNA Helicase

Helicases are a class of enzymes vital to all organisms. Their main function is to unpackage an organism's genes. They are motor proteins that move directionally along a nucleic acid phosphodiester backbone, separating two annealed nucleic acid strands (i.e., DNA, RNA, or RNA-DNA hybrid) using energy derived from ATP hydrolysis. There are many helicases resulting from the great variety of processes in which strand separation must be catalyzed. Approximately 1% of eukaryotic genes code for helicases. The human genome codes for 95 non-redundant helicases: 64 RNA helicases and 31 DNA helicases. Many cellular processes, such as DNA replication, transcription, translation, recombination, DNA repair, and ribosome biogenesis involve the separation of nucleic acid strands that necessitates the use of helicases.

## Function

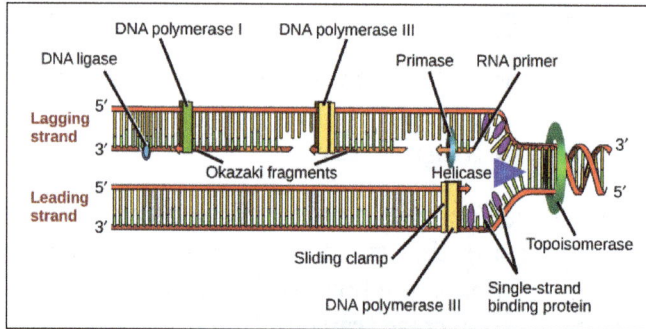

Helicase action in DNA replication

Helicases are often used to separate strands of a DNA double helix or a self-annealed RNA molecule using the energy from ATP hydrolysis, a process characterized by the breaking of hydrogen bonds between annealed nucleotide bases. They also function to remove nucleic acid-associated proteins and catalyze homologous DNA recombination. Metabolic processes of RNA such as translation, transcription, ribosome biogenesis, RNA splicing, RNA transport, RNA editing, and RNA degradation are all facilitated by helicases. Helicases move incrementally along one nucleic acid strand of the duplex with a directionality and processivity specific to each particular enzyme.

Helicases adopt different structures and oligomerization states. Whereas DnaB-like helicases unwind DNA as ring-shaped hexamers, other enzymes have been shown to be active as monomers or dimers. Studies have shown that helicases may act passively, waiting for uncatalyzed unwinding to take place and then translocating between displaced strands, or can play an active role in catalyzing strand separation using the energy generated in ATP hydrolysis. In the latter case, the helicase acts comparably to an active motor, unwinding and translocating along its substrate as a direct result of its ATPase activity. Helicases may process much faster *in vivo* than *in vitro* due to the presence of accessory proteins that aid in the destabilization of the fork junction.

## Activation Barrier in Helicase Activity

Enzymatic helicase action, such as unwinding nucleic acids is achieved through the lowering of the activation barrier ($B$) of each specific action. The activation barrier is a result of various factors, and can be defined using the following equation,

$$B = N(\Delta G_{bp} - G_{int} - G_f)$$

where,

$N$ = number of unwound base pairs (bps),

$\Delta G_{bp}$ = free energy of base pair formation,

$G_{int}$ = reduction of free energy due to helicase, and

$G_f$ = reduction of free energy due to unzipping forces.

Factors that contribute to the height of the activation barrier include: specific nucleic acid sequence of the molecule involved, the number of base pairs involved, tension present on the replication fork, and destabilization forces.

## Active and Passive Helicases

The size of the activation barrier to overcome by the helicase contributes to its classification as an active or passive helicase. In passive helicases, a significant activation barrier exists (defined as $B > k_B T$, where $k_B$ is Boltzmann's constant and $T$ is temperature of the system). Because of this significant activation barrier, its unwinding progression is affected largely by the sequence of nucleic acids within the molecule to unwind, and the presence of destabilization forces acting on the replication fork. Certain nucleic acid combinations will decrease unwinding rates (i.e. guanine and cytosine), while various destabilizing forces can increase the unwinding rate. In passive systems, the rate of unwinding ($V_{un}$) is less than the rate of translocation ($V_{trans}$) (translocation along the single-strand nucleic acid, ssNA). Another way to view the passive helicase is its reliance on the transient unraveling of the base pairs at the replication fork to determine its rate of unwinding.

In active helicases, $B < k_B T$, where the system lacks a significant barrier, as the helicase is able to destabilize the nucleic acids, unwinding the double-helix at a constant rate, regardless of the nucleic acid sequence. In active helicases, $V_{un}$ is approximately equal to $V_{trans}$. Another way to view the active helicase is its ability to directly destabilize the replication fork to promote unwinding.

Active helicases show similar behavior when acting on both double-strand nucleic acids, dsNA, or ssNA, in regards to the rates of unwinding and rates of translocation, where in both systems $V_{un}$ and $V_{trans}$ are approximately equal.

These two categories of helicases may also be modelled as mechanisms. In such models the passive helicases are conceptualized as Brownian ratchets, driven by thermal fluctuations and subsequent anisotropic gradients across the DNA lattice. The active helicases, in contrast, are conceptualized as stepping motors – also known as powerstroke motors – utilizing either a conformational "inch worm" or a hand-over-hand "walking" mechanism to progress. Depending upon the organism, such helix-traversing progress can occur at rotational speeds in the range of 5,000 to 10,000 R.P.M.

## Structural Features

The common function of helicases accounts for the fact that they display a certain degree of amino acid sequence homology; they all possess sequence motifs located in the interior of their primary structure, involved in ATP binding, ATP hydrolysis and translocation along the nucleic acid substrate. The variable portion of the amino acid sequence is related to the specific features of each helicase.

The presence of these helicase motifs allows putative helicase activity to be attributed to a given protein, but does not necessarily confirm it as an active helicase. Conserved motifs do, however, support an evolutionary homology among enzymes. Based on these helicase motifs, a number of helicase superfamilies have been distinguished.

## Superfamilies

Helicases are classified in 6 groups (superfamilies) based on their shared sequence motifs. Helicases not forming a ring structure are in superfamilies 1 and 2, and ring-forming helicases form part of superfamilies 3 to 6. Helicases are also classified as α or β depending on if they work with single or double-strand DNA; α helicases work with single-strand DNA and β helicases work with double-strand DNA. They are also classified by translocation polarity. If translocation occurs 3'-5' the helicase is type A; if translocation occurs 5'-3' it is type B.

- Superfamily 1 (SF1): This superfamily can be further subdivided into SF1A and SF1B helicases. In this group helicases can have either 3'-5' (SF1A subfamily) or 5'-3'(SF1B subfamily) translocation polarity. The most known SF1A helicases are Rep and UvrD in gram-negative bacteria and PcrA helicase from gram-positive bacteria. The most known Helicases in the SF1B group are RecD and Dda helicases.

- Superfamily 2 (SF2): This is the largest group of helicases that are involved in varied cellular processes. They are characterized by the presence of nine conserved motifs: Q, I, Ia, Ib, and II through VI. This group is mainly composed of DEAD-box RNA helicases. Some other helicases included in SF2 are the RecQ-like family and the Snf2-like enzymes. Most of the SF2 helicases are type A with a few exceptions such as the XPD family.

- Superfamily 3 (SF3): Superfamily 3 consists of helicases encoded mainly by small DNA viruses and some large nucleocytoplasmic DNA viruses. They have a 3'-5' translocation directionality, meaning that they are all type A helicases. The most known SF3 helicase is the papilloma virus E1 helicase.

- Superfamily 4 (SF4): All SF4 family helicases have a type B polarity (5'-3'). The most studied SF4 helicase is gp4 from bacteriophage T7.

- Superfamily 5 (SF5): Rho proteins conform the SF5 group.

- Superfamily 6 (SF6): They contain the core AAA+ that is not included in the SF3 classification. Some proteins in the SF6 group are: mini chromosome maintenance MCM, RuvB, RuvA, and RuvC.

# Helicase Disorders and Diseases

## ATRX Helicase Mutations

The *ATRX* gene encodes the ATP-dependent helicase, ATRX (also known as XH2 and XNP) of the SNF2 subgroup family, that is thought to be responsible for functions such as chromatin remodeling, gene regulation, and DNA methylation. These functions assist in prevention of apoptosis, resulting in cortical size regulation, as well as a contribution to the survival of hippocampal and cortical structures, affecting memory and learning. This helicase is located on the X chromosome (Xq13.1-q21.1), in the pericentromeric heterochromatin and binds to heterochromatin protein 1. Studies have shown that ATRX plays a role in rDNA methylation and is essential for embryonic development. Mutations have been found throughout the *ATRX* protein, with over 90% of them being located in the zinc finger and helicase domains. Mutations of ATRX can result in X-linked-alpha-thalassaemia-mental retardation (ATR-X syndrome).

Various types of mutations found in ATRX have been found to be associated with ATR-X, including most commonly single-base missense mutations, as well as nonsense, frameshift, and deletion mutations. Characteristics of ATR-X include: microcephaly, skeletal and facial abnormalities, mental retardation, genital abnormalities, seizures, limited language use and ability, and alpha-thalassemia. The phenotype seen in ATR-X suggests that the mutation of ATRX gene causes the downregulation of gene expression, such as the alpha-globin genes. It is still unknown what causes the expression of the various characteristics of ATR-X in different patients.

## XPD Helicase Point Mutations

XPD (Xeroderma pigmentosum factor D, also known as protein ERCC2) is a 5'-3', Superfamily II, ATP-dependent helicase containing iron-sulphur cluster domains. Inherited point mutations in XPD helicase have been shown to be associated with accelerated aging disorders such as Cockayne syndrome (CS) and trichothiodystrophy (TTD). Cockayne syndrome and trichothiodystrophy are both developmental disorders involving sensitivity to UV light and premature aging, and Cockayne syndrome exhibits severe mental retardation from the time of birth. The XPD helicase mutation has also been implicated in xeroderma pigmentosa (XP), a disorder characterized by sensitivity to UV light and resulting in a several 1000-fold increase in the development of skin cancer.

XPD is an essential component of the TFIIH complex, a transcription and repair factor in the cell. As part of this complex, it facilitates nucleotide excision repair by unwinding DNA. TFIIH assists in repairing damaged DNA such as sun damage. A mutation in the XPD helicase that helps form this complex and contributes to its function causes the sensitivity to sunlight seen in all three diseases, as well as the increased risk of cancer seen in XP and premature aging seen in trichothiodystrophy and Cockayne syndrome.

XPD helicase mutations leading to trichothiodystrophy are found throughout the protein in various locations involved in protein-protein interactions. This mutation results in an unstable protein due to its inability to form stabilizing interactions with other proteins at the points of mutations. This, in turn, destabilizes the entire TFIIH complex, which leads to defects with transcription and repair mechanisms of the cell.

It has been suggested that XPD helicase mutations leading to Cockayne syndrome could be the result of mutations within XPD, causing rigidity of the protein and subsequent inability to switch from repair functions to transcription functions due to a "locking" in repair mode. This could cause the helicase to cut DNA segments meant for transcription. Although current evidence points to a defect in the XPD helicase resulting in a loss of flexibility in the protein in cases of Cockayne syndrome, it is still unclear how this protein structure leads to the symptoms described in Cockayne syndrome.

In xeroderma pigmentosa, the XPD helicase mutation exists at the site of ATP or DNA binding. This results in a structurally functional helicase able to facilitate transcription, however it inhibits its function in unwinding DNA and DNA repair. The lack of cell's ability to repair mutations, such as those caused by sun damage, is the cause of the high cancer rate in xeroderma pigmentosa patients.

## RecQ Family Mutations

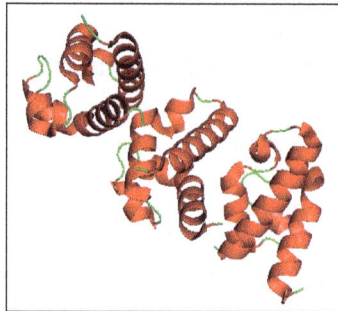

RecQ helicase

RecQ helicases (3'-5') belong to the Superfamily II group of helicases, which help to maintain stability of the genome and suppress inappropriate recombination. Deficiencies and mutations in RecQ family helicases display aberrant genetic recombination and DNA replication, which leads to chromosomal instability and an overall decreased ability to proliferate. Mutations in RecQ family helicases BLM, RECQL4, and WRN, which play a role in regulating homologous recombination, have been shown to result in the autosomal recessive diseases Bloom syndrome (BS), Rothmund-Thomson syndrome (RTS), and Werner syndrome (WS), respectively.

Bloom syndrome is characterized by a predisposition to cancer with early onset, with a mean age-of-onset of 24 years. Cells of Bloom syndrome patients show a high frequency of reciprocal exchange between sister chromatids (SCEs) and excessive chromosomal damage. There is evidence to suggest that BLM plays a role in rescuing disrupted DNA replication at replication forks.

Werner syndrome is a disorder of premature aging, with symptoms including early onset of atherosclerosis and osteoporosis and other age related diseases, a high occurrence of sarcoma, and death often occurring from myocardial infarction or cancer in the 4th to 6th decade of life. Cells of Werner syndrome patients exhibit a reduced reproductive lifespan with chromosomal breaks and translocations, as well as large deletions of chromosomal components, causing genomic instability.

Rothmund-Thomson syndrome, also known as poikiloderma congenitale, is characterized by premature aging, skin and skeletal abnormalities, rash, poikiloderma, juvenile cataracts, and a predisposition to cancers such as osteosarcomas. Chromosomal rearrangements causing genomic instability are found in the cells of Rothmund-Thomson syndrome patients.

## Meiotic Recombination

During meiosis DNA double-strand breaks and other DNA damages in a chromatid are repaired by homologous recombination using either the sister chromatid or a homologous non-sister chromatid as template. This repair can result in a crossover (CO) or, more frequently, a non-crossover (NCO) recombinant. In the yeast *Schizosaccharomyces pombe* the FANCM-family DNA helicase FmI1 directs NCO recombination formation during meiosis. The RecQ-type helicase Rqh1 also directs NCO meiotic recombination. These helicases, through their ability to unwind D-loop intermediates, promote NCO recombination by the process of synthesis-dependent strand annealing.

In the plant *Arabidopsis thaliana*, FANCM helicase promotes NCO and antagonizes the formation of CO recombinants. Another helicase, RECQ4A/B, also independently reduces COs. It was suggested that COs are restricted because of the long term costs of CO recombination, that is, the breaking up of favorable genetic combinations of alleles built up by past natural selection.

## RNA Helicases

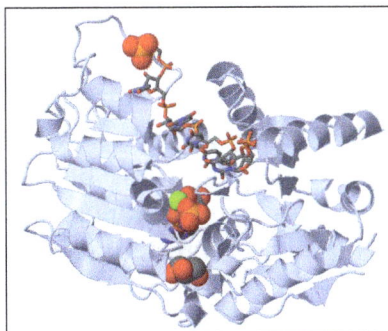

Human DEAD-box RNA helicase

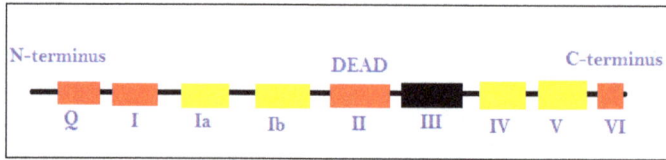

This image represents the different promoter sequences and accessory domains that aid in RNA unwinding (local strand separation). The regions in red are ATP binding domains and the regions in yellow are RNA interaction domains. Specific sequences termed DEAD box proteins are also present that help catalyze reactions in which ATP does not need to be directly hydrolyzed, as long as it binds to the domains on the strand.

RNA helicases are essential for most processes of RNA metabolism such as ribosome biogenesis, pre-mRNA splicing, and translation initiation. They also play an important role in sensing viral RNAs. RNA helicases are involved in the mediation of antiviral immune response because they can identify foreign RNAs in vertebrates. About 80% of all viruses are RNA viruses and they contain their own RNA helicases. Defective RNA helicases have been linked to cancers, infectious diseases and neuro-degenerative disorders. Some neurological disorders associated with defective RNA helicases are: amyotrophic lateral sclerosis, spinal muscular atrophy, spinocerebellar ataxia type-2, Alzheimer disease, and lethal congenital contracture syndrome.

RNA helicases and DNA helicases can be found together in all the helicase superfamilies except for SF6. All the eukaryotic RNA helicases that have been identified up to date are non-ring forming and are part of SF1 and SF2. On the other hand, ring-forming RNA helicases have been found in bacteria and viruses. However, not all RNA helicases exhibit helicase activity as defined by enzymatic function, i.e., proteins of the Swi/Snf family. Although these proteins carry the typical helicase motifs, hydrolize ATP in a nucleic acid-dependent manner, and are built around a helicase core, in general, no unwinding activity is observed.

RNA helicases that do exhibit unwinding activity have been characterized by at least two different mechanisms: canonical duplex unwinding and local strand separation. Canonical duplex unwinding is the stepwise directional separation of a duplex strand, as described above, for DNA unwinding. However, local strand separation occurs by a process wherein the helicase enzyme is loaded at any place along the duplex. This is usually aided by a single-strand region of the RNA, and the loading of the enzyme is accompanied with ATP binding. Once the helicase and ATP are bound, local strand separation occurs, which requires binding of ATP but not the actual process of ATP hydrolysis. Presented with fewer base pairs the duplex then dissociates without further assistance from the enzyme. This mode of unwinding is used by the DEAD/DEAH box helicases.

An RNA helicase database is currently available online that contains a comprehensive list of RNA helicases with information such as sequence, structure, and biochemical and cellular functions.

## Diagnostic Tools for Helicase Measurement

## Measuring and Monitoring Helicase Activity

Various methods are used to measure helicase activity *in vitro*. These methods range from assays that are qualitative (assays that usually entail results that do not involve values or measurements) to quantitative (assays with numerical results that can be utilized in statistical and numerical analysis). In 1982–1983, the first direct biochemical assay was developed for measuring helicase activity. This method was called a "strand displacement assay".

Strand displacement assay involves the radiolabeling of DNA duplexes. Following helicase treatment, the single-strand DNA is visually detected as separate from the double-strand DNA by non-denaturing PAGE electrophoresis. Following detection of the single-strand DNA, the amount of radioactive tag that is on the single-strand DNA is quantified to give a numerical value for the amount of double-strand DNA unwinding.

The strand displacement assay is acceptable for qualitative analysis, its inability to display results for more than a single time point, its time consumption, and its dependence on radioactive compounds for labeling warranted the need for development of diagnostics that can monitor helicase activity in real time.

Other methods were later developed that incorporated some, if not all of the following: high-throughput mechanics, the use of non-radioactive nucleotide labeling, faster reaction time/less time consumption, real-time monitoring of helicase activity (using kinetic measurement instead of endpoint/single point analysis). These methodologies include: "a rapid quench flow method, fluorescence-based assays, filtration assays, a scintillation proximity assay, a time resolved fluorescence resonance energy transfer assay, an assay based on flashplate technology, homogenous time-resolved fluorescence quenching assays, and electrochemiluminescence-based helicase assays". With the use of specialized mathematical equations, some of these assays can be utilized to determine how many base paired nucleotides a helicase can break per hydrolysis of 1 ATP molecule.

Commercially available diagnostic kits are also available. One such kit is the "Trupoint" diagnostic assay from PerkinElmer, Inc. This assay is a time-resolved fluorescence quenching assay that utilizes the PerkinEmer "SignalClimb" technology that is based on two labels that bind in close proximity to one another but on opposite DNA strands. One label is a fluorescent lanthanide chelate, which serves as the label that is monitored through an adequate 96/384 well plate reader. The other label is an organic quencher molecule. The basis of this assay is the "quenching" or repressing of the lanthanide chelate signal by the organic quencher molecule when the two are in close proximity – as they would be when the DNA duplex is in its native state. Upon helicase activity on the duplex, the quencher and lanthanide labels get separated as the DNA is unwound. This loss in proximity negates the quenchers ability to repress the lanthanide signal, causing a detectable increase in fluorescence that is representative of the amount of unwound DNA and can be used as a quantifiable measurement of helicase activity. The execution and use of single-molecule fluorescence imaging techniques, focusing on methods that include optical trapping in conjunction with epifluorescent imaging, and also surface immobilization in conjunction with total internal reflection fluorescence visualization. Combined with microchannel flow cells and microfluidic control, allow individual fluorescently labeled protein and DNA molecules to be imaged and tracked, affording measurement of DNA unwinding and translocation at single-molecule resolution.

## Determining Helicase Polarity

Helicase polarity, which is also deemed "directionality", is defined as the direction (characterized as 5'→3' or 3'→5') of helicase movement on the DNA/RNA single-strand along which it is moving. This determination of polarity is vital in f.ex. determining whether the tested helicase attaches to the DNA leading strand, or the DNA lagging strand. To characterize this helicase feature, a partially duplex DNA is used as the substrate that has a central single-strand DNA region with different lengths of duplex regions of DNA (one short region that runs 5'→3' and one longer region that runs 3'→5') on both sides of this region. Once the helicase is added to that central single-strand region, the polarity is determined by characterization on the newly formed single-strand DNA.

## Amylase

Amylase is a digestive enzyme that aids in the breakdown of carbohydrates by breaking the bonds between sugar molecules in polysaccharides through a hydrolysis reaction. It can be found in animals, plants, and bacteria.

Amylase can be classified into three types: alpha-amylase, beta-amylase, and gamma-amylase. The three types differ in how they hydrolyze the polysaccharide bonds. Neither beta nor gamma-amylase is found in animal tissue. This information sheet will focus on alpha-amylase, which is an important enzyme in digestive and metabolic processes.

## Function in the Body

Amylase is critical in the digestion of starch into sugars to make them available energy sources for the body. Amylase is found in two primary places within the human body, and the two types are classified according to where they are found. Salivary Amylase is a component of saliva, and breaks starch into glucose and dextrin. It hydrolyzes the bonds between long-chain polysaccharides found in food, breaking compounds such as glycogen and starch into their useful monomers, glucose and maltose. Pancreatic Amylase is added to the small intestine to further digest starches; amylase is denatured in the acidic stomach. Amylase is also present in blood where it digests dead white blood cells.

## Types of Amylases

Enzymes belonging to amylases, endoamylases and exoamylases, are able to hydrolyse starch. These enzymes are classified according to the manner in which the glycosidic bond is attacked. The starch degrading enzymes are found in the numerous glycoside hydrolase families 13(GH- 13 families).

## α- Amylase (EC 3.2.1.1)

Endoamylases are able to cleave α,1-4 glycosidic bonds present in the inner part (endo-) of the amylose or amylopectin chain. α-amylase is a well-known endoamylase. It is found in a wide variety of microorganisms, belonging to the Archaea as well as the bacteria. The end products of α-amylase action are oligosaccharides with varying length with α-configuration and α-limit dextrins, which constitute branched oligosaccharides. α- amylases are often divided into two categories according to the degree of hydrolysis of the substrate. Saccharifying α- amylases hydrolyze 50 to 60% and liquefying α-amylases cleave about 30 to 40% of the glycosidic linkages of starch. The α-amylases are calcium metalloenzymes, completely unable to function in the absence of calcium., α-amylase breaks down long-chain carbohydrates by acting at random locations along the starch chain, ultimately yielding maltotriose and maltose from amylose, or maltose, glucose and "limit dextrin" from amylopectin. α-amylase tends to be faster- acting than β-amylase because it can act anywhere on the substrate. In human physiology, both the salivary and pancreatic amylases are α-Amylases and are also found in plants (adequately), fungi (ascomycetes and basidiomycetes) and bacteria (Bacillus).

## β- Amylase (E.C. 3.2.1.2)

Enzymes belonging to the second group, the exoamylases, either exclusively cleave α,1-4 glycosidic bonds such as β-amylase or cleave both α,1-4 and α,1-6 glycosidic bonds like amyloglucosidase or glucoamylase (E.C. 3.2.1.3) and α-glucosidase (E.C. 3.2.1.20). Exoamylases act on the external glucose residues of amylose or amylopectin and thus produce only glucose (glucoamylase and α-glucosidase), or maltose and β-limit dextrin. β-amylase and glucoamylase also convert the anomeric configuration of the liberated maltose from α to β. Glucoamylase and α-glucosidase differ in their substrate preference, α-glucosidase acts best on short maltooligosaccharides and liberates glucose with α-configuration while glucoamylase hydrolyzes long-chain polysaccharides best. β-amylases and glucoamylases have also been found in a large variety of microorganisms.

## γ-Amylase (EC 3.2.1.3)

γ-amylase cleaves α(1-6) glycosidic linkages, in addition to cleaving the last α (1-4) glycosidic linkages at the non-reducing end of amylose and amylopectin, yielding glucose.

Unlike the other forms of amylase, γ-amylase is most efficient in acidic environments and has an optimum pH of 3.

## Sources of Alfa Amylase

## Plant Sources

Amylases are widespread in animals, fungi, plants, and are also found in the unicellular eukaryotes, bacteria and archaea. Though plants and animals produce amylases, enzymes from microbial sources are generally used in industrial processes. This is due to a number of factors including productivity, thermostability of the enzyme as well as ease of cultivating microorganisms. The major advantages of the enzymatic route are the selectivity with its associated high yield and exclusivity toward the desired product. Bacteria used in commercial production are the Bacillus spp. Others, such as Escherichia spp, Pseudomonas, Proteus, Serratia and Rhizobium also yield appreciable quantity of the enzyme. Aspergillus, Rhizopus, Mucor, Neurospora, Penicillium and Candida are some of the fungi that also produce extracellular amylases of commercial value. Plant sources had not been considered with enough significance as the source of these enzymes yet. The utilization of agriculture waste materials serves two functions: reduction in pollution and upgrading of these materials. Agricultural wastes are being used for both liquid and solid fermentation to reduce the cost of fermentation media. These wastes consist of carbon and nitrogen sources necessary for the growth and metabolism of organisms. These nutrient sources included pearl millet starch, orange waste, potato, corn, tapioca, wheat and rice as flours were used for α-amylase production. α-Amylases are being produced commercially in bulk from microorganisms and represent about 25-33% of the world enzyme market. They had numerous applications including liquefaction of starch in the traditional beverages, baking and textile industry for desizing of fabrics. Moreover, they have been applied in paper manufacture, medical fields as digestives and as detergent additives.

## Animal Sources

Ptyalin, a salivary α-amylase (α-1,4-α-D-glucan-4-glucanohydrolase; E.C. 3.2.1.1) is one of the most important enzymes in saliva. The enzyme was first described in saliva by Leuchs in 1831. It consists of two families of isoenzymes, of which one set is glycosylated and the other contains no carbohydrate. The molecular weight of the glycosylated form is about 57 kDa; that of the non-glycosylated form is about 54 kDa. Salivary amylases accounts for 40% to 50% of the total salivary protein and most of the enzyme being synthesized in the parotid gland (80% of the total). It is a calcium-containing metalloenzyme that hydrolyzes the α-1,4 linkages of starch to glucose and maltose. It is known to be mainly involved in the initiation of the digestion of starch in the oral cavity. However, Salivary α-amylase has also been shown to have an important bacterial interactive function.

## Microbial Sources

In spite of the wide distribution of amylases, microbial sources, mainly fungal and bacterial amylases, are used for the industrial production due to advantages such as cost effectiveness, consistency, less time and ease of process modification and optimization. Fungal amylases have been widely used for the preparation of oriental foods. Among bacteria, Bacillus sp. is widely used for thermostable α-amylase production to meet industrial needs. B. subtilis, B. stearothermophilus, B. licheniformis and B. amyloliquefaciens are known to be good producers of α-amylase and these have been widely used for commercial production of the enzyme for various applications. Similarly, filamentous fungi have been widely used for the production of amylases for centuries. As these moulds are known to be prolific producers of extracellular proteins, they are widely exploited for the production of different enzymes including α-amylase.

## Fungal Amylases

Fungi belonging to the genus Aspergillus have been most commonly employed for the production of α-amylase. With the development of genetic engineering, Bacillus subtilis is becoming an increasingly attractive host for cloning. The advantages of B. subtilis such as high secretion level and non-pathogenic safe (GRAS-generally recognized as safe) status for non- antibiotics strains have made it suitable for the production of heterologous enzymes. Most reports about fungi that produce α-amylase have been limited to a few species of mesophilic fungi, and attempts have been made to specify the cultural conditions and to select superior strains of the fungus to produce on a commercial scale. Fungal sources are confined to terrestrial isolates, mostly to Aspergillus and Penicillium. The Aspergillus species produce a large variety of extracellular enzymes, and amylases are the ones with most significant industrial importance. Filamentous fungi, such as Aspergillus oryzae and Aspergillus niger, produce considerable quantities of enzymes that are used extensively in the industry. A. oryzae has received increased attention as a favorable host for the production of heterologous proteins because of its ability to secrete a vast amount of high value proteins and industrial enzymes, e.g. α-amylase. Aspergillus oryzae has been largely used in the production of food such as soy sauce, organic acid such as citric and acetic acids and commercial enzymes including α-amylase. Aspergillus niger has important hydrolytic capacities in the α-amylase production and, due to its tolerance of acidity (pH < 3), it allows the avoidance of bacterial contamination. Filamentous fungi are suitable microorganisms for solid state fermentation (SSF), especially because their morphology allows them to colonize and penetrate the solid substrate. The fungal α-amylases are preferred over other microbial sources due to their more accepted GRAS status. The thermophilic fungus Thermomyces lanuginosus is an excellent producer of thrmostable amylase purified the α-amylase, proving its thermostability.

## Bacterial Amylases

The production of microbial amylases from bacteria is dependent on the type of strain, composition of medium, method of cultivation, cell growth, nutrient requirements, incubation period, pH, temperature, metal ions and thermostability. In fact, such industrially important microorganisms found within the genus Bacillus, can be exploited commercially due to their rapid growth rate leading to short fermentation cycles, capacity to secrete proteins into the extracellular medium and safe handling.

α-Amylase can be produced by different species of microorganisms, but for commercial applications α-amylase is mainly derived from the genus Bacillus. α-Amylases produced from Bacillus licheniformis, Bacillus stearothermophilus, and Bacillus amyloliquefaciens find potential application in a number of industrial processes such as in food, fermentation, textiles and paper industries. Thermostability is a desired characteristic of most of the industrial enzymes. Thermostable enzymes isolated from thermophilic organisms have found a number of commercial applications because of their stability. As enzymatic liquefaction and saccharification of starch are performed at high temperatures (100–110 °C), thermostable amylolytic enzymes have been currently investigated to improve industrial processes of starch degradation and are of great interest for the production of valuable products like glucose, crystalline dextrose, dextrose syrup, maltose and maltodextrins. Bacillus subtilis, Bacillus stearothermophilus, Bacillus licheniformis, and Bacillus amyloliquefaciens are known to be good producers of thermostable α-amylase, and these have been widely used for commercial production of the enzyme for various applications. Thermostable α-amylases have been reported from several bacterial strains and have been produced using SmF as well as SSF. However, the use of SSF has been found to be more advantageous than SmF and allows a cheaper production of enzymes. The production of α-amylase by SSF is limited to the genus Bacillus, and B. subtilis, B. polymyxia, B. mesentericus, B. vulgarus, B. megaterium and B. licheniformis have been used for α-amylase production in SSF. Currently, thermostable amylases of Bacillus stearothermophilus or Bacillus licheniformis are being used in starch processing industries. Enzymes produced by some halophilic microorganisms have optimal activity at high salinities and could therefore be used in many harsh industrial processes where the concentrated salt solutions used would otherwise inhibit many enzymatic conversions. In addition, most halobacterial enzymes are considerably thermotolerant and remain stable at room temperature over long periods. Halophilic amylases have been characterized from halophilic bacteria such as Chromohalobacter sp. Halobacillus sp. Haloarcula hispanica, Halomonas meridian and Bacillus dipsosauri.

Bacillus is endowed to produce thermostable α-amylase and also large quantities of other enzymes. Indeed, 60% of commercially available enzymes are obtained from different species of Bacillus i.e. B. subtilis, B. stearothermophilus, B. licheniformis and B. amyloliquefaciens. Some Bacillus strains produce enzyme in the exponential phase, whereas some others in the mid stationary phase. Though, different Bacillus species have similar growth patterns and enzyme profiles, but their optimized conditions vary, depending upon the strain. Some properties exhibited by different bactria are shown in following table:

Table: Bacterial α-amylases & their characteristics

| Microorganisms | Mode of Fermentation | pH optima | Temp. optima | Mol. Wt. (kDa) | Inhibitors |
|---|---|---|---|---|---|
| *Chromohalobacter* sp. TVSP 101 | SSF | 7.0 - 9.0 | 65 °C | 72 | - |
| *Haloarcula hispânica* | | 6.5 | 50 °C | 43.3 | EDTA |
| *Bacillus* sp. I-3 | SmF | 7.0 | 70 °C | - | EDTA, HgCl2 |
| *Bacillus* sp. PN5 | SmF | 10 | 60 °C | - | NH4Cl |
| *BacillussubtilisDM-03* | SSF | 6.0–10 | 50 °C | - | - |
| *Bacillus subtilis* KCC103 | SmF | 6.5 | 37 °C | - | - |
| *Bacillus* sp. KCA102 | | 7.1 | 57.5 °C | - | - |
| *Bacillus subtilis JS- 2004* | SmF | 7.0 | 50 °C | - | $Co^{2+}$ $Cu^{2+}$ $Hg^{2+}$ $Mg^{2+}$ $Zn^{2+}$ $Ni^{2+}$ $Fe^{2+}$ and $Mn^{2+}$ |
| *Bacillus subtilis* | SmF | 7.0 | 135 °C | - | - |
| *Bacillus caldolyticus* DSM405 | SmF | 5.0-6.0 | 70 °C | - | - |
| *Bacillus* sp. Ferdowsicous | | 4.5 | 70 °C | 53 | $Hg^{2+}Zn^{2+}$ EDTA |
| *Halomonas meridian* | SmF | 7.0 | 37 °C | - | Glu |
| *Geobacillus thermoleovorans* | - | 7.0 | 70 °C | - | - |

## Mode of Action of Alfa Amylase

In general, it is believed that α-amylases are endo-acting amylases which hydrolyze α- (1-4) glycosidic bonds of the starch polymers internally. Several models for amylase action pattern have been proposed, such as the random action and the multiple attack action. Random action has also been referred to as a single attack or multi-chain attack action. In the former, the polymer molecule is successively hydrolysed completely before dissociation of the enzyme- substrate complex. While, in the latter, only one bond is hydrolysed per effective encounter. The multiple attack action is an intermediate between the single-chain and the multi-chain action where the enzyme cleaves several glycosidic bonds successively after the first (random) hydrolytic attack before dissociating from the substrate. It is observed that the multiple attack action is generally an accepted concept to explain the differences in action pattern of amylases. However, most of the endoamylases have a low to very low level of multiple attack action increased with temperature to a degree depending on the amylase itself.

## Structure of Alfa Amylase

### Molecular Weight

Despite wide difference of microbial α-amylases characters, their molecular weights are usually in the same range 40-70 kDa while the highest molecular weight of α-amylases, 210 kDa, for Chloroflexus aurantiacus. Whereas, 10 kDa of Bacillus caldolyticus α-amylase was reported to be the lowest value. This molecular weight may be raised due to glycosylation as in the case of Thermoactiomyces vulgaris α-amylase that reach 140 kDa. In contrast, proteolysis may lead to decrease in the molecular weight. For example, α-amylase of T. vulganis 94-2A (AmyTV1) is a protein of 53 kDa and smaller peptides of 33 and 18 kDa that have been shown to be products of limited AmyTV1 proteolysis.

### Molecular Structure

α-amylases from different organisms share about 30% amino acid sequence identity and all belong to the same Glycosyl Hydrolase family 13 (GH-13 family of protein). The three dimensional (3D) structures of α-amylases have revealed monomeric, calcium-containing enzymes, with a single polypeptide chain folded into three domains (A-C). The most conserved domain in α-amylase family enzymes, the A-domain, consists of a highly symmetrical fold of eight parallel β-strands arranged in a barrel encircled by eight α-helices. The highly conserved amino acid residues of the α-amylase family involved in catalysis and substrate binding are located in loops at the C-termini of β-strands in this domain. This is typical to all enzymes belonging to the α/β −barrel protein family. α-amylases have a B-domain that protrudes between β-sheet number 3 and α-helix number. 3. It ranges from 44 to 133 amino acid residues and plays a role in substrate or Ca binding. The sequence of this domain varies most; in Bacillus, α-amylases it is relatively long and folds into a more complex structure of β-strands, whereas in barley α-amylase there is an irregularly structured domain of 64 residues. All known α-amylases, with a few exceptions, contain a conserved $Ca^{+2}$ binding site which is located at the interface between domains A and B. In addition, α-amylase produced by Bacillus thermooleovorans was found to contain a chloride ion binding site in their active site, which has been shown to enhance the catalytic efficiency of the enzyme, presumably by elevating the pKa of the hydrogen-donating residue in the active site.

α-amylases has a domain C which is relatively conserved and folds into an antiparallel β barrel. The orientation of domain C relative to domain A varies depending on the type and source of amylase. The function of this domain is unknown. Structural studies have confirmed that the active sites of glycosyl hydrolases are composed of multiple binding sites, or subsites, for the sugar units of polymeric substrates. The open active site cleft is formed between domains A and B, so that residues from domain B participate in substrate binding. The substrate binding sites are commonly lined with aromatic residues (Phe, Trp and Tyr) which make hydrophobic stacking interactions with the sugar rings. In addition, the active sites contain many residues which form hydrogen bonds to the substrate either directly or via water molecules. In Taka-amylase A, the first examined protein α-amylase by X-ray crystallography, three acidic residues, i.e., one glutamic and two aspartic acids were found at the centre of the active site, and subsequent mutational studies have shown that these residues are essential for catalysis. The glutamic acid residue is now believed to be the proton donor, while the first of the two conserved aspartic acids appearing in the amino acid sequence of an α- amylase family member is thought to act as the nucleophile. The role of the second aspartic acid is less certain, but it has been suggested to involved in stabilising the oxocarbenium ion-like transition state and also in maintaining the glutamic acid in the correct state of protonation for activity. These residues occur near the ends of strands 3, 4, 5 and 7 of the a/β -barrel and are found in four short sequences, long-recognised as being conserved in α-amylase family enzymes.

## Applications of Alfa Amylase

Starch is a major storage product of many economically important crops such as wheat, rice, maize, tapioca, and potato. A large-scale starch processing industry has emerged in the last century. In the past decades, we have seen a shift from the acid hydrolysis of starch to the use of starch-converting enzymes in the production of maltodextrin, modified starches, or glucose and fructose syrups. Currently, these enzymes comprise about 30 % of the world's enzyme production. Besides the use in starch hydrolysis, starch-converting enzymes are also used in a number of other industrial applications, such as laundry and porcelain detergents or as anti- staling agents in baking. A number of these starch-converting enzymes belong to a single family: the alpha amylase family or family13 glycosyl hydrolases. This group of enzymes share a number of common characteristics such as α (β/α)8 barrel structure, the hydrolysis or formation of glycosidic bonds in the a conformation, and a number of conserved amino acid residues in the active site. As many as 21 different reaction and product specificities are found in this family.

## Bread and Chapatti Industry

The quantities, taste, aroma and porosity of the bread are improved by using the enzyme in the flour. More than 70 % bread in U.S.A, Russia and European countries contain alpha amylase. Amylases play important role in bakery products. For decades, enzymes such as malt and fungal alpha-amylases have been used in bread-making. The significance of enzymes is likely to raise as consumers insist more natural products free of chemical additives. For example, enzymes can be employed to replace potassium bromate, a chemical additive that has been prohibited in a number of countries. The dough for bread, rolls, buns and similar products consists of flour, water, yeast, salt and possibly other ingredients such as sugar and fat. Flour consists of gluten, starch, non-starch polysaccharides, lipids and trace amounts of minerals. As soon as the dough is made, the yeast starts to work on the fermentable sugars, transforming them into alcohol and carbon dioxide, which makes the dough rise. The major component of wheat flour is starch. Amylases can degrade starch and produce small dextrins for the yeast to act upon. The alpha-amylases degrade the damaged starch in wheat flour into small dextrins, which allows yeast to work continuously during dough fermentation, proofing and the early stage of baking. The result is improved bread volume and crumb texture. In addition, the small oligosaccharides and sugars such as glucose and maltose produced by these enzymes enhance the Maillard reactions responsible for the browning of the crust and the development of an attractive baked flavor.

## Textile Industry

Textile industries are extensively using alpha amylases to hydrolyze and solubilize the starch, which then wash out of the cloth for increasing the stiffness of the finished products. Fabrics are sized with starch. Alpha amylase is used as desizing agent for removing starch from the grey cloth before its further processing in bleaching and dyeing. Many garments especially the ubiquitions 'Jean' are desized after mashing. The desired fabrics are finally laundered and rinsed.

## Sugar and Glucose Industries

Alpha amylase plays a very important role in the production of starch conversion products of low fermentability. The presence of starch and other polysaccharides in sugar cane creates problem throughout the sugar manufacturing which is minimized or eliminated by the action of alpha amylase. The high quality products depends upon the efficiency of the enzyme which lead to low production, costs for the starch processor has increased. Many industries used alpha amylases for the production of glucose. Enzyme hydrolyzed the starch and converted it into glucose. They hydrolyze α-1,4 glucosidic linkage in the starch polymer in a random manner to yield glucose and maltose. There-fore, alpha amylase is extensively used in many industries for the production of glucose and also used in water-soluble dextrin.

## Alcohol Industry

Alpha amylases convert starch in to fermentable sugars. Starches such as grain;po-tatoes etc. are used as a raw material that helps to manufacture ethyl alcohol. In the presence of amylases, the starch is first converted in to fermentable sugars. The use of bacterial enzyme partly replaces malt in brewing industry, thus making the process more economically significant. Alpha amylase can also carries out the reactions of alco-holysis by using methanol as a substrate.

## Paper Industry

Starch paste when used as a mounting adhesive modified with additives such as pro-tein glue or alum, frequently, causes damage to paper as a result of its embrittlement. Starch digesting enzymes, e.g. alpha amylase, in immersion or as a gel poultice are applied to facilitate its removal. Alpha amylase hydrolyzed the raw starch that is used for sizing and coating the paper instead of expensive chemically modified starches. So, starch is extensively used for some paper size press publications.

## Detergent, Building Product and Feed Industries

In detergent industries, the enzyme alpha amylase plays a vital role. It is widely used for improvement of detergency of laundry bleach composition and bleaching with out color darkening. The addition of enzyme stabilizes the bleach agent and preserves effectiveness of the bleach in laundry detergent bar composition. Modified starch is used in the manufacture of gypsum board for dry wall construction. Enzyme modified the starch for the industry use. Many starches or barely material are present in the feed. So, the nutritional value of the feed can be improved by the addition of alpha amylase.

## Chocolate Industry

Amylases are treated with cocoa slurries to produce chocolate syrup, in which choco-late starch is dextrinizing and thus syrup does not become thick. Cocoa flavored syrups having a high cocoa content and excellent stability and flow properties at room em-perature may be produced by using an amylolytic enzyme and a sufficient proportion of Dutch process cocoa to provide a syrup pH of 5.5 to 7.5. The syrup is made by alter-nate addition of cocoa and sweetener to sufficient water to achieve a solids content of about 58 to 65 weight percent, adding an amylolytic enzyme, heating to a temperature of about 175 -185 °F for at least 10 to 15 min, raising the temperature to about 200 °F. and cooling. The stabilized cocoa flavored syrups may be added at room temperature to conventional non-acid confection mixes for use in the production of quiescently frozen chocolate flavored confections.

## Fuel Alcohol Production

Ethanol is the most utilized liquid biofuel. For the ethanol production, starch is the most used substrate due to its low price and easily available raw material in most regions of the world. In this production, starch has to be solubilized and then submitted to two enzymatic steps in order to obtain fermentable sugars. The bioconversion of starch into ethanol involves liquefaction and saccharification, where starch is converted into sugar using an amylolytic microorganism or enzymes such as α-amylase, followed by fermentation, where sugar is converted into ethanol using an ethanol fermenting microorganism such as yeast Saccharomyces cerevisiae. The production of ethanol by yeast fermentation plays an important role in the economy of Brazil. In order to obtain a new yeast strain that can directly produce ethanol from starch without the need for a separate saccharifying process, protoplast fusion was performed between the amylolytic yeast Saccharomyces fibuligera and S. cerevisiae. Among bacteria,α-amylase obtained from thermoresistant bacteria like Bacillus licheniformis or from engineered strains of Escherichia coli or Bacillus subtilis is used during the first step of hydrolysis of starch suspensions.

## Treatment of Starch Processing Waste Water (SPW)

Starch is also present in waste produced from food processing plants. Starch waste causes pollution problems. Biotechnological treatment of food processing waste water can produce valuable products such as microbial biomass protein and also purifies the effluent.

## Analysis in Medicinal and Clinical Chemistry

With the advent of new frontiers in biotechnology, the spectrum of amylase applications has expanded into many other fields, such as clinical, medicinal and analytical chemistry. There are several processes in the medicinal and clinical areas that involve the application of amylases. The application of a liquid stable reagent, based on α-amylase for the Ciba Corning Express clinical chemistry system has been described. A process for the detection of higher oligosaccharides, which involved the application of amylase was also developed. This method was claimed to be more efficient than the silver nitrate test. Biosensors with an electrolyte isolator semiconductor capacitor (EIS-CAP) transducer for process monitoring were also developed.

## Other Applications

- An inhibitor of alpha-amylase called phaseolamin has been tested as a potential diet aid.

- A higher than normal concentration of amylases may reflect one of several medical conditions, including acute inflammation of the pancreas (concurrently with the more specific lipase), but also perforated peptic ulcer, torsion of an ovarian cyst, strangulation ileus, macroamylasemia and mumps. Amylase may be measured in other body fluids,including urine and peritoneal fluid.

- In molecular biology, the presence of amylase can serve as an additional method of selecting for successful integration of a reporter construct in addition to antibiotic resistance. As reporter genes are flanked by homologous regions of the structural gene for amylase, successful integration will disrupt the amylase gene and prevent starch degradation, which is easily detectible through iodine staining.

# Ribozymes

A ribozyme is a ribonucleic acid (RNA) enzyme that catalyzes a chemical reaction. The ribozyme catalyses specific reactions in a similar way to that of protein enzymes.

Also called catalytic RNA, ribozymes are found in the ribosome where they join amino acids together to form protein chains. Ribozymes also play a role in other vital reactions such as RNA splicing, transfer RNA biosynthesis, and viral replication.

The first ribozyme was discovered in the early 1980s and led to researchers demonstrating that RNA functions both as a genetic material and as a biological catalyst. This contributed to the worldwide hypothesis that RNA may have played a crucial role in the evolution of self-replicating systems. This is referred to as the RNA World Hypothesis and today, many scientists believe that ribozymes are remnants of an ancient world that existed before the evolution of proteins. It is thought that RNAs used to catalyse functions such as cleavage, replication and RNA molecule ligation before proteins evolved and took over these catalytic functions, which they could perform in a more efficient and versatile way.

Ribozymes have been extensively investigated by researchers to try and determine their exact structure and function. Scientists have developed synthetic ribozymes in the laboratory that are able to catalyze their own synthesis under specific conditions. One example is the RNA polymerase ribozyme. Using mutagenesis and selection, scientists have managed to develop and improve variants of the Round-18 polymerase ribozyme. The best variant so far is called B6.61, which can add up to 20 nucleotides to a primer template over a period of 24 hours. After 24 hours, the hydrolysis of its phosphodiester bonds causes the ribozyme to decompose.

Ribozymes may also play an important role in therapeutic areas, acting as molecules that can tailor specific RNA sequences, serving as biosensors and providing a useful tool in areas such as gene research and functional genomics. For example, strands of circular ribozymes celled viriods have been discovered and these can have a devastating effect on plants. The viriods replicate by making copies of themselves based on their own genome and their catalytic properties enable them to undergo self-cleavage and send fragments off to colonize and harm areas of a plant by proliferating and using the genetic material that the plant itself needs. Researchers have now identified a site in these viriods that enables them to self-cleave. The site is less than 30 nucleotides in length and has three stems that form a central loop which is referred to as a "hammerhead." This structure cleaves very specific RNA sequences to release viable RNA daughter strands. Now, hammerheads of just 19 nucleotides in length have been synthesized that act as highly specific catalysts. Similar ribozymes are also being made that could be used to break up RNA viruses and RNA that is required for the transcription and translation of DNA that contains mutations.

## General Characteristics

Catalytic RNAs are broadly separated into two classes based on their size and reaction mechanisms. The large catalytic RNAs consist of RNase P, and the group I and group II introns. These molecules range in size from a few hundred nucleotides to around 3000. They catalyze reactions that generate reaction intermediates and products with 3′ hydroxyls and 5′ phosphates. The small catalytic RNAs include the hammerhead, the hairpin (or paperclip), hepatitis delta and VS RNA. These molecules range in size from ~35 to ~155 nucleotides. They use the 2′ hydroxyl of the ribose sugar as a nucleophile, and they generate products with a 2′,3′-cyclic phosphate and a 5′ hydroxyl. The relationship between the size and the reaction mechanism of these molecules has raised intriguing questions about their origins and evolution. It may be that the reaction mechanism and the size of the large ribozymes are needed to bring often very distal elements of the substrate into close proximity. The small, self-cleaving, RNAs are not faced with this constraint and perhaps this permitted them to evolve smaller catalytic centers. It remains possible, however, that the relationship between the size and reaction mechanism is simply fortuitous.

Reactions catalyzed by RNAs. a: The reaction mechanism of the large catalytic RNAs. The nucleophile BOH is the 3′ hydroxyl of a guanosine cofactor in the first step of group I splicing, the 2′ hydroxyl of a nucleotide, generally A, within the intron in the first step of group II splicing, and a water molecule in the RNase P-catalyzed reaction. b: The reaction mechanism of the small catalytic RNAs. The reaction is most likely initiated by the activation of the 2′ hydroxyl on the ribose located at the scissile bond, and it results in a product with a 2′,3′-cyclic phosphate. The nature of the activation is still poorly understood, but it most likely involves a coordinated metal hydroxide. Notice in both (a) and (b) that the net number of bonds is conserved throughout the reaction.

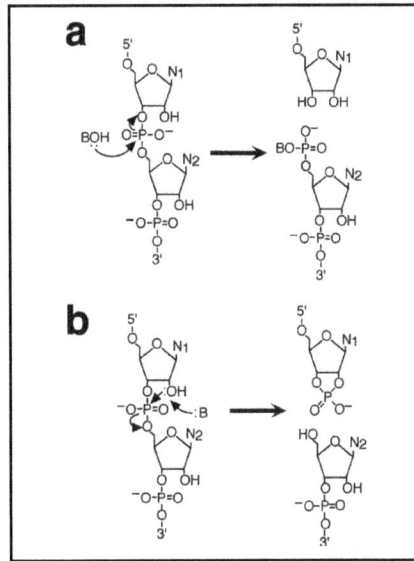

With one exception, all these RNAs catalyze reactions that modify themselves. Hence, they cannot be considered true enzymes or catalysts. The exception is RNase P, which processes the 5' end of tRNA precursors. It is the only known example of a naturally occurring RNA-based enzyme. However, all these molecules can be converted, with some clever engineering, into true RNA enzymes that modify other RNAs in trans without becoming altered themselves.

Ribozymes increase reaction rates by up to $10^{11}$-fold and have reaction efficiencies, kcat/Km, up to $10^8$ $M^{-1}$ $min^{-1}$, which is in the range for diffusion-controlled duplex formation between oligonucleotides. While impressive, the rate enhancements provided by ribozymes are still $\sim 10^3$-fold less than those provided by protein enzymes catalyzing comparable reactions. Moreover, ribozymes cannot compare with proteins as multiple-turnover enzymes, mostly because product release is so slow that the catalytic site of the ribozyme is easily saturated. This may be an inherent limitation of RNA enzymes, but it could also reflect evolutionary constraints, since ribozymes generally catalyze intramolecular, single-turnover, reactions in nature. An exhaustive comparison of the enzymatic mechanistics of protein and RNA enzymes has recently been made.

All known ribozymes have an absolute requirement for a divalent cation, which is generally $Mg^{2+}$. Some, notably within the large catalytic RNAs, require divalent cations for proper assembly of the tertiary structures as well. On this basis, catalytic RNAs are considered to be metalloenzymes, and a general two-metal-ion reaction mechanism has been proposed for the large catalytic RNAs, based on analogy with the properties of protein metalloenzymes. The role of divalent cations for the small catalytic RNAs is less clear, but they are generally considered to be essential for catalysis.

## Properties of Introns

Introns, or intervening sequences (IVS), are nonencoding sequences that interrupt the coding, exon, sequences. These introns must be removed, at the RNA level, in order for the gene to be expressed functionally. There are five major categories of introns and splicing mechanisms. These consist of nuclear tRNA introns, archaeal introns, nuclear mRNA introns, and the group I and group II introns. Of these introns, some members of the group I and group II are clearly capable of catalyzing their own excision, in vitro, in an RNA-catalyzed fashion. However, genetic analyses have revealed protein factors that are often essential for group I and group II splicing in vivo. The nuclear tRNA and archaeal introns clearly require protein factors (endonuclease and ligase). Nuclear mRNA introns may represent an intermediate state. These introns are spliced within large (40–60S) ribonucleoprotein particles (RNP), which consist of a number of small nuclear RNAs (snRNAs) and proteins. However, it is increasingly thought that it is the snRNAs that are the chemical engines of the complex as well as the determinants of the splice sites.

One of the oddities of group I and group II introns, which makes the term 'intron' something of an oxymoron, is that they sometimes encode proteins that are required in vivo either for splicing activity or for intron mobility. The former are intron maturases, which can be highly specific for the intron that encodes it. The latter are endonucleases with reverse-transcriptase properties. These proteins are involved in intron 'homing', where crosses in yeast (the best studied genetically) result in the transfer of the intron into the intronless allele of the gene. There is also intron transposition (or 'retrohoming'), where the intron is inserted into other alleles. The widespread, but scattered, distribution of introns – especially introns with high sequence similarity occupying the same location in different organisms – suggests the possibility for horizontal transfer of introns; thus these introns may act somewhat akin to infectious agents. In group I introns, mobility is apparently catalyzed by proteins alone and a functional (active) intron is not required. In group II, the intron itself is essential for activity and the intron RNA actually becomes inserted into the double-stranded DNA substrate.

## Group I Introns

Group I introns range in size from a few hundred nucleotides to around 3000. They are abundant in fungal and plant mitochondria, but they are also found in nuclear rRNA genes, chloroplast DNA (ctDNA), bacteriophage, eukaryotic viruses, and in the tRNA of ctDNA and eubacteria. In short, they are widely found except in higher eukaryotes. The various group I introns have little sequence similarity, but they are characterized by four short conserved sequence elements, called P, Q, R, and S. P can always partially base-pair with Q, and R can always partially base-pair with S. In addition, all group I introns can fold into distinctive, phylogenetically conserved, secondary structures consisting of 10 paired segments. Additional sequences, including large open reading frames (ORF), and structures are often found, but they do not disrupt the catalytic core, which consists of P3, P4, P6 and P7. However, their presence is used to subdivide the introns into various subgroups.

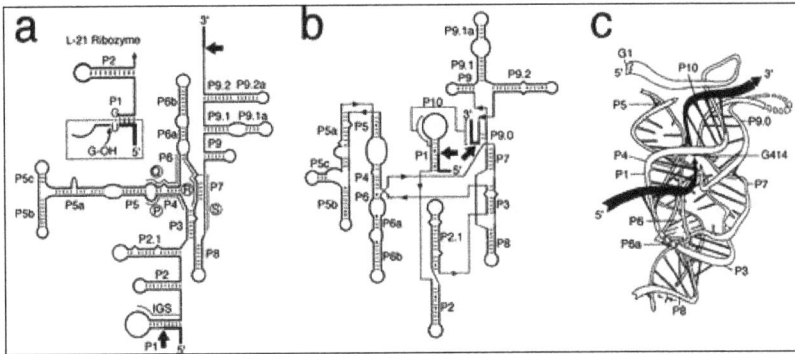

The structure of group I introns. a: The 'classical' depiction of the secondary structure of the rRNA intron from T. thermophila, showing the phylogenetically conserved
sequences P, Q, R, and S (circled) and the conserved pairings P1–P9. Nonconserved
elements are designated by additional numbers or letters (e.g., P9.1 and P5c). The IGS
sequence, which forms P1 and P10, is as indicated. The exon sequences are indicated as
heavy lines and the splice sites by the arrows. The insert shows the 5' end of the L-21
ribozyme, which is used in various reactions in trans, with a bound substrate. b: The
new secondary structure depiction that more accurately represents the relationships of
the different structural elements. The heavy lines are exon sequences and the very light
ones are used to bridge the different elements, which are very close in tertiary space. c:
A computer-generated model of the catalytic core of the Tetrahymena intron. The exon
sequences are shown in solid and the intron as outline. This figure shows the intron
after the first reaction, where the nonencoded guanosine, G1, is attached to the 5' end
of the intron and the 3' hydroxyl of the 5' exon is posed to make a nucleophilic attack
at the 3' splice site.

The secondary structure was originally deduced from computer modeling based on phylogenetic comparisons and, in the case of the Tetrahymena intron, on limited structural
probing. Additional evidence for this structure comes from enzymatic and chemical
probing studies, additional phylogenetic analyses and from physical approaches. These
studies have confirmed the overall structure and revealed other important interactions,
which have resulted in a new secondary structure depiction that more accurately reflects the spatial relationship of the different elements.

The exon sequences are oriented relative to the catalytic core by base pairings with
an intron sequence called the internal guide sequence (IGS), which constitutes P1 and
P10 in the structure. An additional pairing, called P9.0, further aligns the intron-3'
exon splice site. It was so named because it occurs between P8 and P9 in the secondary
structure, the latter being named before the interaction was discovered. The 5' exon
contains a highly conserved uridine on its 3' end, which forms a functionally important
U·G base pair with the IGS, and the intron contains a conserved guanosine on its 3'
end. Finally, there is a specific binding site for the guanosine cofactor that initiates the
reaction. This site involves a conserved G-C base pair in P7.

Michel and Westhof have made a very clever, computer-generated, three-dimensional model of the catalytic core based on careful phylogenetic comparisons. This model brings all the relevant elements into proximity, and it has provided a sound basis for additional studies. A crystal structure was solved to 2.8 Å for a 160-nucleotide fragment of the Tetrahymena intron, consisting of P4–P6. While this structure contains less than half of the catalytic core of the enzyme and it tells us little about the reaction mechanism, it has revealed several exciting features. These include a tetraloop, GNRA (N is any base and R is purine), docking with its receptor, several adenosine platforms, which are derived from adjacent adenosines forming a pseudo-base pair within the helix, and a ribose zipper, which involves a network of hydrogen bonds between the 2'-OH groups of ribose and acceptor groups of bases within the shallow groove of helices.

Recently, a 5.0-Å crystal structure has been solved for a 247-nucleotide-long fragment of the Tetrahymena intron, consisting of helices P3–P9. This new structure shows the previously characterized P4–P6 domain (helices P4, P5, and P6) largely unchanged and the P3–P9 domain (helices P3, P7, P8 and P9) wrapped around it. The close packing of the two domains creates a shallow cleft into which the P1 helix, containing the 5' splice site, could fit. The structure also creates a particularly tight binding site in the P7 helix for the guanosine cofactor that initiates the splicing reaction. Unfortunately, the resolution is too low to visualize details within the structure, and the crystal structure is still missing the P1–P2 domain containing the splice site, so little can be determined about the reaction mechanism. However, it is clear that the computer generated model for the catalytic core is largely consistent with this crystal structure. A triumph for computer modeling of RNA structures.

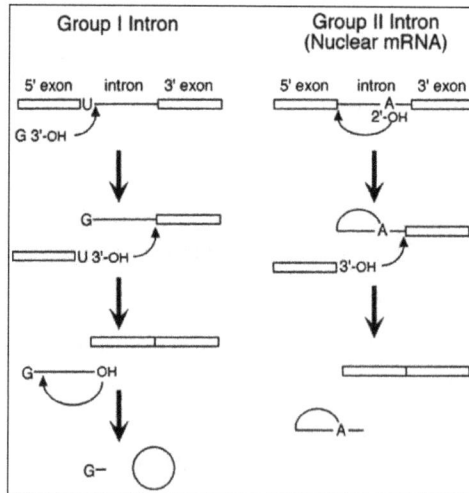

The splicing reaction of group I introns was first worked out for Tetrahymena thermophila, and it will form the basis for my further discussion. Nevertheless, all characterized group I introns follow essentially the same pathway. The Tetrahymena intron is excised from the precursor rRNA by a two-step transesterification reaction. The reaction is initiated by the nucleophilic attack of the 3' hydroxyl of a guanosine cofactor at the 5' splice site. The exon-intron phosphodiester bond is cleaved and the guanosine forms a 3',5'-phosphodiester bond at the 5' end of the intron. The now free 3' hydroxyl of the 5' exon then makes a nucleophilic attack at the 3' splice site to form the ligated exons and release the intron with the nonencoded guanosine.

Splicing mechanisms of group I and group II introns. In both cases, a series of trans-esterification reactions are used to excise the intron and ligate the exons. The net number of bonds remains the same throughout. The reaction is initiated by a guanosine cofactor in group I introns and by an internal adenosine in group II. The splicing reaction of nuclear pre-mRNAs follows the same pathway as group II introns, but it occurs on a large ribonucleoprotein complex.

Nucleotides close to the 5' end of the intron are realigned on the IGS and the highly conserved 3' terminal guanosine of the intron makes a nucleophilic attack at a phosphodiester bond between nucleotides 15 and 16 or between nucleotides 19 and 20 within the intron in a reaction that is analogous to the first step of splicing. The intron is circularized (C-15 or C-19) and a small fragment containing the nonencoded guanosine is released. The circular product is also found in vivo, and its formation presumably helps to drive the reaction to completion. Although the circularization reaction is not universally conserved among group I introns, it is very common. Each step is essentially the forward or reverse of the same reaction, and the total number of phosphodiester bonds is conserved. Each step is fully reversible, and no external energy source is needed. The intron can fully reintegrate into rRNAs both in vitro and in vivo. The phosphodiester bonds are inverted during the reaction, which is consistent with an SN2, in-line, reaction mechanism.

The circular product is normally considered the end product of the reaction, but the released intron retains catalytic activity and substrate specificity. For T. thermophila, a linear form of the intron lacking the first 21 and the last five nucleotides (L-21 ScaI) will catalyze a wide range of reactions on substrates added in trans. These include sequence-specific endonuclease, nucleotidyltransferase, ligase and phosphatase activities. Substrate specificity is changed by altering the IGS sequence. Because the reactions are catalyzed using the 3' hydroxyl, they will work on both RNA and DNA substrates, although the latter has a much lower binding affinity and hence less activity. Since the L-21 ScaI RNA is unaltered in the reaction, it is considered a true RNA-based enzyme.

The properties of group I introns lend themselves to a number of different applications. The L-21 ScaI version of the Tetrahymena intron has been commercially sold as an RNA restriction enzyme. Circularly permutated precursor RNA, which contains end-to-end fused exons inserted within the middle of the intron sequence, is used to generate circular exon molecules by an 'inverse' splicing reaction. A variant of this technique is used to generate circular, trans-cleaving, HDV ribozymes in vitro and in vivo, which are more resistant to nucleases. A trans-splicing reaction can repair a truncated lacZ transcript in Escherichia coli and in the cytoplasm of mammalian cells. A trans-splicing intron can also change mutant β-globin transcripts, in sickle cell anemia, into mRNAs coding for antisickling γ-globin in human erythroid lineage cells.

## Group II Introns

Group II introns range in size from several hundred to around 2500 nucleotides. Although they are much less widely distributed than group I introns, they are found in fungal and plant mitochondria, in chloroplasts of plants, in algae, in eubacteria and especially in the chloroplasts of the protist Euglena gracilis. Most are present in mRNAs, but a few also occur in tRNA and rRNA genes. In contrast to the extensively analyzed group I introns, much less is known about group II introns. This is partly because of their more limited distribution but also because very few of them are found to be self-splicing in vitro. Those that are autocatalytic require reaction conditions that are far from physiological (e.g., 100 mM $MgCl_2$, 500 mM $(NH_4)_2SO_4$ and 45 °C).

The secondary structure of group II introns was originally deduced from phylogenetic comparisons and computer modeling. It is normally depicted as six helical domains (I–VI) radiating as spokes from a central wheel. On the basis of the structural features, group II introns are divided into two major subclasses, although some introns fit into neither class. Some contain long ORFs. Important structural features have been difficult to elucidate because of their poor reactivity in vitro and the difficulty of working with organelle-specific molecules. However, it appears that only domains I and V are indispensable. In addition, domain V contains most of the relatively few phylogenetically conserved nucleotides found in group II introns and it might constitute the reaction center. Domain VI contains the highly conserved adenosine that is generally used to initiate the splicing reaction. Except for domains I and V, the domains can be modified or deleted and the intron will retain some catalytic activity. Optional ORFs are frequently located in the loop of domain IV.

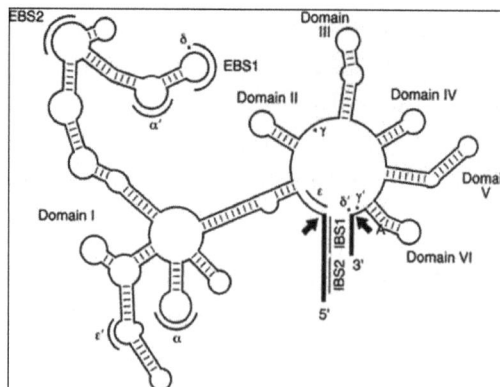

The secondary structure of group II introns. This cartoon is a generalization from a number of different introns. The exon sequences are indicated as heavy lines and the splice sites by the arrows. The characteristics of the central wheel with the radiating domains is conserved, but the characteristics of the individual domains vary considerably. Tertiary interactions are formed between IBS1-EBS1, IBS2-EBS2, α-α',?-?', δ-δ' and γ-γ'; the connections between these elements are for clarity. The conserved adenosine used to initiate the splicing reaction is indicated in domain VI. Additional tertiary interactions have been identified, which vary with the intron.

Degenerated forms of group II introns are found in plant chloroplasts and mitochondria that often lack recognizable cognates of the various domains. These are called group III introns they may require factors in trans for activity or be assembled from parts of group II introns. These degenerate introns provide a feasible pathway between group II and nuclear mRNA splicing, where more and more of the role of the intron is supplanted by trans-acting factors. This possibility is made more plausible by the observation that some group II and group III introns occur within other group II or group III introns (called twintrons;) and that others are discontinuous. Segments of these latter introns are transcribed within two or even three separate molecules from distant regions of the genome and the exons are assembled by a trans-splicing reaction.

The 5' exon is aligned by interactions between two intron binding sequences (IBS1 and IBS2), located near the 3' end of the 5' exon, and two exon binding sequences (EBS1 and EBS2), which are located in domain I of the intron. The 5' splice site is further defined by?-?' interactions. These interactions are important both before and after 5' cleavage. The 3' splice site has multiple determinants, comprising δ-δ', γ-γ', and other as yet unidentified interactions. As yet, no tertiary model has been proposed.

Unlike group I introns, where the first and second steps are thought of as forward and reverse steps of the same reaction, implying a single catalytic site, group II introns use two different nucleophiles (2' and 3' hydroxyls) and both steps show the same stereospecificity. This suggests that there are two independent reaction centers or that there is a single site that switches substrates.

The splicing reaction is generally initiated by the nucleophilic attack of the 2' hydroxyl of a highly conserved adenosine in domain VI to form a distinctive structure, called a lariat, containing 3'-5' and 2'-5' phosphodiester bonds at the adenosine branch site. The now free 3' hydroxyl of the 5' exon then makes a nucleophilic attack at the 3' splice site to form the ligated exons and release the intron, still as a lariat. The reaction can also be initiated, both in vitro and in vivo, by the nucleophilic attack of water, although this is not a typical route. In this case, the intron is released as a linear molecule. As mentioned above, these reactions can also be catalyzed in trans from separate transcripts. The splicing mechanism and the lariat product are reminiscent of those found in nuclear mRNA splicing, and it provides an additional evolutionary link between the two splicing reactions.

The applications of group II introns are more limited than those of group I. Nevertheless, it can catalyze the cleavage of ligated precursors using the energy of a phosphoanhydride bond, ligate RNA to DNA, and cleave single-stranded DNA substrates. It is also used to circularize exon sequences,. However, the fact that the intron RNA becomes inserted into double-stranded DNA during intron homing suggests that other applications will soon be found.

## RNase P

RNase P is a ubiquitous enzyme in all characterized organisms that processes the 5′ termini of tRNA precursors. In eubacteria, RNase P exists as a ribonucleoprotein complex, consisting of a large RNA of about 350–400 nucleotides (M1 RNA in E. coli) and a small basic protein of ~14 kDa (C5 in E. coli). The basic protein is essential for activity in vivo, however the RNase P RNA, by itself, can catalyze the reaction in vitro. This reaction requires high salt concentrations (e.g., 1 M $K^+$ or $NH4^+$ and 10 mM MgCl2;), which suggests that the basic protein component acts only as an electrostatic shield to promote binding between the RNA enzyme and the RNA substrate. However, the protein also affects cleavage-site specificity and turnover, so its full role in the reaction is still unclear. RNase P from E. coli will also process other substrates that partially resemble tRNAs (e.g., 4.5S RNA), but these reactions appear to be relatively minor. RNase P RNA is the only characterized ribozyme that, unmodified, acts in trans on multiple substrates, and hence it is considered the only true, naturally occurring, RNA enzyme.

RNase P from the nuclei and mitochondria of eukaryotes also exist as ribonucleoprotein complexes, although they generally have much higher protein contents (ca. 50–70% versus ca. 10% for eubacteria). Moreover, while the RNA component is inevitably found to be essential, the RNase P RNA from eukaryotes has never been shown to have catalytic activity. Archaebacteria have many properties that make them more similar to the eukaryotes than to the eubacteria, and they likewise are not known to have RNA-alone catalytic activity.

There is evidence that another ribonuclease in eukaryotes, RNase MRP, is closely related, and perhaps homologous (i.e., evolutionarily related), to RNase P. It is a ribonucleoprotein complex that participates in nucleolar pre-rRNA processing. The RNA component can fold into similar secondary structures as RNase P RNA and, in yeast, it shares common protein components. However, like the eukaryotic RNase P, the RNase MRP RNA has not been shown to have catalytic activity by itself.

Although there is little sequence conservation, all the eubacterial RNase P RNAs can be folded into similar, although not identical, secondary structures on the basis of comparative sequence analyses. The E. coli M1 RNA consists of 18 paired helices (P1–P18; Fig.). The Bacillus subtilis M1 RNA analog is more diversified from E. coli than many of the other eubacteria. It folds into a similar structure, but P6, P13, P14, P16 and P17 are missing and it contains extra helices P5.1, P10.1, P15.1, and P19. These two RNAs are the most extensively studied, but through a careful comparative analysis of the different eubacteria, it is possible to derive a common core structure consisting of helices P1–P5, P7–P12 and P15. However, the activity of this 'minimal' structure has not yet been demonstrated.

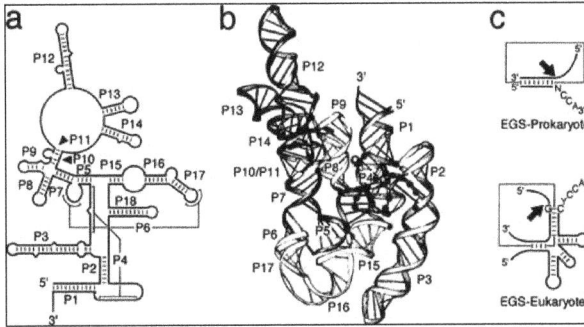

Characteristics of RNase P RNA. a: The proposed secondary structure of M1 RNA, the RNA component of RNase P from E. coli. b: The computer-modeled tertiary structure of (a). The black spheres represent invariant nucleotides at the catalytic site and conserved nucleotides in the T-loop recognition site. Note that these conserved residues are clustered close to each other in what forms the catalytic core of the molecule. c: Synthetic substrates for RNase P from prokaryotes and eukaryotes. The EGS, which is used to define the target specificity, is shown bound to the substrate (boxed) and the cleavage site is indicated by an arrow.

Native RNase P RNAs differ from the phylogenetic minimum by having extra stems and stem-loop structures. While probably not needed for catalysis, these structures nevertheless lower the ionic strength requirements and enhance their thermal stability. Such elements may be redundant in the sense that any single one can be deleted or modified without significantly altering the activity, but it is not possible to simultaneously delete or modify all of them. The RNase P RNA from eukaryotes and archaebacteria have little sequence similarity to their eubacterial counterparts, but they can often be folded into a universally conserved core structure as well. Missing RNA elements from the structures may be supplanted by the protein component(s), but this has not been characterized.

Three-dimensional models of the E. coli M1 RNA have been made by Westhof and Altman and by Harris et al., and both groups have recently refined these models. Three-dimensional models for the B. subtilis RNase P RNA are also available. These are computer-derived models based on data obtained from phylogenetic comparisons, mutational analyses, chemical probing and from crosslinking studies. Both groups' models have similar overall structures and both provide a pocket or cleft into which the tRNA substrate will fit. However, many of the specific interactions vary; thus these models are expected to undergo continual refinement as additional experimental data become available.

RNase P uses water as a nucleophile to cleave the phosphodiester bond. The exact mechanism by which the tRNA precursor is bound is still unclear. The 3' half of the acceptor stem is thought to function as an external guide sequence (EGS), but it does not uniquely define the cleavage site. The primary sequence does not seem to be important nor does any single element uniquely define the cleavage site. Instead, recognition could result from several redundant factors, including the distance along the coaxially stacked T stem-loop and acceptor stem, the 3' terminal CCA sequence, the EGS alignment, and a conserved guanosine 3' to the cleavage site. Thus, recognition is largely, if not entirely, based on tertiary interactions with the substrate.

From studies using small substrates, it is possible to design synthetic EGSs that can target any RNAs for cleavage by RNase P in vitro or in vivo. The basis for this is that the EGS binds to the target RNA and makes it look like a tRNA substrate. In eubacteria, the minimal requirement is that it forms a short stem with a free NCCA. The target sequence can be virtually anything. In eukaryotes, the EGS-substrate complex must more closely resemble a tRNA. This reaction works in vitro, in bacteria and in human cells. A variant of this technique is to add the appropriate EGS (here called an internal guide sequence or IGS) to the 3' end of the RNase P RNA. This increases the efficiency of the reaction, and it is used to inactivate thymidine kinase mRNA from herpes simplex virus in cell lines. Yet, while the therapeutic potential of RNase P has been demonstrated, it has not been widely used in therapeutic applications.

## General Properties of Small Catalytic RNAs

Self-cleaving RNAs are generally found in small ($\sim$ 220 to $\sim$ 460 nucleotides long) RNA pathogens of plants known as viroids, virusoids and linear satellite viruses. However, they are also found within satellite RNAs of salamanders, Neurospora, and within another pathogenic satellite virus found in man. The viroid and satellite RNAs are generally replicated by an RNA-dependent rolling-circle mechanism, and the catalytic domains are thought to process the linear concatemers that are generated into unit-length progeny.

The linear, unit-length progenies produced during replication in vivo are subsequently ligated to form closed-circular molecules that are used in the next round of rolling-circle replication. It is reasonable to expect that this is catalyzed by the ribozyme as well since, mechanistically, it represents the reverse of the cleavage reaction, and it would be analogous to the splicing reaction carried out by the group I and group II introns. However, in vitro only the hairpin ribozyme shows significant ligation activity. Protein factors may be involved in vivo or other, as yet unidentified, RNA elements may be required.

Four motifs are characterized, and they all catalyze reactions that generate products with 2', 3'-cyclic phosphates and 5' hydroxyls. As with the previous ribozymes, they all require a divalent metal ion, normally $Mg^{2+}$, for activity.

The intramolecular self-cleaving activity is converted into a trans-cleaving activity by making the 'substrate' and 'ribozyme' into separate molecules. However, the 'substrate' remains an integral part of the structure of the active ribozyme, and hence the ribozyme is often simply defined as the unmodified portion of the molecule and the substrate is the cleaved portion. Hence, the 'ribozyme' may consist of different sequence elements, depending on the construct. The catalytic domains of these ribozymes are small and relatively well characterized, and they are more widely used in therapeutic applications. Each has characteristics that confer specific advantages and disadvantages as therapeutic agents. The in vitro and ex vivo activity of cis-cleaving forms of three of these self-cleaving ribozymes (hammerhead, hairpin and HDV) have been compared.

## Hammerhead Ribozyme

The hammerhead ribozyme is probably the most extensively studied of all the ribozymes, and it is the motif most commonly found in the viroids and satellite RNAs. Currently 16 hammerhead motifs are known in the plus and minus strands of these plant pathogens. Three other hammerhead motifs are found in the satellite 2 RNAs from the salamanders, Triturus vulgaris, Ambystoma talpoideum and Amphiuma tridactylum. This ribozyme was so named because its Australian discoverers found the secondary structure, as originally drawn, to be reminiscent of the head of a hammerhead shark. It is the smallest of the naturally occurring self-cleaving RNAs, at 40–50 nucleotides in length.

Analyses of the hammerhead ribozyme are voluminous. It consists of three helical regions, which are variable, and three single-stranded regions that contain most of the highly conserved nucleotides. The length of the helical arms can be quite variable, and helix II can be reduced to two base pairs. Mutating any of the conserved residues markedly reduces activity; consequently, important functional groups are often identified by incorporating synthetic nucleotide analogs into the RNA.

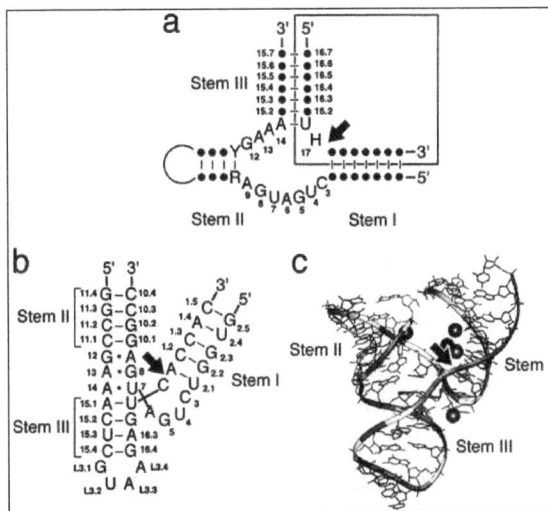

Characteristics of the hammerhead ribozyme. a: The secondary structure of the hammerhead ribozyme showing the conserved sequence and structure. The dots represent nucleotides that can be anything, Y is a pyrimidine, R is a purine and H is any nucle-otide except guanosine. The arrow indicates the self-cleavage site. The boxed region shows the portion that is normally the substrate in trans-cleaving versions of the ribozyme. The numbering is based on standardized nomenclature. b: A new secondary structure drawing that more accurately reflects the spatial relationships of the different elements. This structure differs from (a) in that it shows a loop in stem III rather than in stem II. c: The solved crystal structure of the sequence shown in (b). This is an RNA-only structure infused with $Mg^{2+}$, shown as spheres. Only the $Mg^{2+}$ ion close to the scissile bond is generally accepted as being functionally relevant.

Cleavage occurs after an NUH triplet, where N is any nucleotide, and H is any nucle-otide except guanosine. The most effective triplet is GUC, but other triplet combinations will work nearly as well; their relative activities have been compared, although the ordering can vary depending on the method of analysis. The reaction mechanism is extensively studied, and it appears to involve a metal-coordinated hydroxide, which probably directly activates the 2' hydroxyl. The reaction products are consistent with an SN2 (in-line) reaction mechanism; this was suggested by an inversion of the phosphate at the scissile linkage.

The hammerhead ribozyme was the first catalytic RNA for which the complete X-ray crystal structure was solved. There are now a number of such structures available with resolutions ranging from 2.6 Å to 3.1 Å, as well as a structure derived from fluorescence resonance energy transfer. The molecule has a Y shape, with stem I and stem II at the arms and stem III at the base. Recently, the crystallized structure of a reaction intermediate was determined using a tallo-5'-C-methyl-ribose-modified ribozyme that is kinetically blocked for the final cleavage reaction. This latter structure is more compatible with an SN2 reaction mechanism. This is in contrast to the previously solved structures that showed the ribozyme in a ground state that was incompatible with such a mechanism. A second $Mg^{2+}$ binding site has potentially been identified in another crystal structure that could be involved in stabilizing the pentacoordinated phosphate transition state. Moreover, there are other incompatibilities between the experimental data and the crystalline structures.

The hammerhead ribozyme is divided into separate 'ribozyme' and 'substrate' in several ways, but the one shown in figure is the most commonly used because most of the conserved residues are contained within the ribozyme rather than in the substrate. The hybridizing arms are varied to optimize ribozyme activity and substrate specificity. Normally hybridizing arms of six or seven base pairs are considered optimal, but for variable arms it is better to have a long stem III and short stem I than the reverse. Often, a tetraloop (frequently GAAA) is used for loop II and stem II is GC-rich to further increase the stability of the stem-loop. The malleability of the hammerhead ribozyme makes it the most commonly used ribozyme for in vivo studies, and there are many successful examples. A trans-cleaving hammerhead ribozyme is approved for phase II clinical trials against HIV-1 by Ribozyme Pharmaceuticals (RPI).

## Hairpin Ribozyme

The hairpin ribozyme is found in three pathogenic, plant, satellite viruses, although the one found in the satellite virus associated with tobacco ring spot virus (sTRSV) is the best characterized. It consists of four stem regions that, when lined up coaxially, somewhat resemble a hairpin; interestingly, it was also originally named 'paperclip,' which may, in fact, better represent its overall three-dimensional shape. It consists of two noncontiguous sequences of 50 and 14 nucleotides within the minus strand of sTRSV. The secondary structure was determined based on computer-aided modeling, limited phylogenetic comparisons, mutational analyses and by in vitro selection.

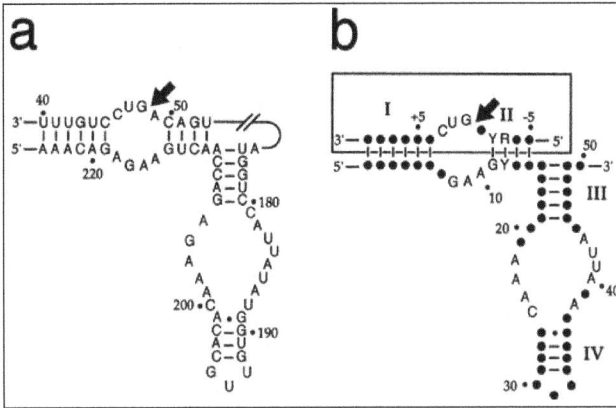

Secondary structure of the hairpin ribozyme. a: The minus strand of sTRSV (numbering is that of the full-length virus). The arrow shows the cleavage site. b: The consensus sequence and structure, where dots are any nucleotide, Y is a pyrimidine and R is a purine. The boxed region represents the portion that is normally the substrate in trans-cleaving reactions. The substrate is numbered relative to the cleavage site and the 'ribozyme' relative to the 5' end.

The other hairpin ribozymes are found in the satellite viruses of arabis mosaic virus (sARMV) and chicory yellow mottle virus (sCYMV), and they mostly differ from the sequence shown in Fig. by nucleotide changes within the helical regions that maintain the structure as shown. Indeed, the mutational and in vitro selection analyses show that the helical regions are structural elements that can largely be changed, as long as the integrity of the helices is maintained. Most of the conserved nucleotides occur within the single-stranded regions. The guanosine 3' to the cleavage site is essential, but altering the other conserved positions can dramatically reduce the activity as well. The bulged region between helices III and IV contain a conserved motif, called a UV-loop motif, that is found in a diverse group of RNAs, including viroids, 5S rRNA and the sarcin-ricin loop of 28S rRNA. However, the role this motif plays in catalysis is still unknown.

Recently, a computer-generated tertiary model was made that was based on preexisting structural data and on the spatial distance of tolerated, interdomain, aryl-disulfide crosslinks. Additional information was also obtained by Walker et al. using FRET data. The current model shows helix I coaxially stacked on helix II and helix IV coaxially stacked on helix III. These two extended helices are then bent so that helix II and helix III, and helix I and helix IV are positioned side by side. This places the two highly conserved bulged regions in proximity, and they could thus form the catalytic core. However, this tertiary model is still preliminary and additional data are required before the details of the catalytic site are known.

A major advantage of the hairpin ribozyme lies in its ability to catalyze both cleavage and ligation reactions efficiently in vitro; this has greatly facilitated in vitro selection experiments because new substrates, with the appropriate PCR primer sites, are easily generated. The RNA-catalyzed ligation reaction is also thought to be relevant in vivo, in that the RNA can both cleave the linear multimers generated during rolling-circle replication and ligate them to form the circular RNA progeny. However, as with the other catalytic motifs used in viroid replication, the cleavage-ligation reaction must be carefully regulated in vivo to prevent inappropriate cleavage, or ligation, of the resulting progeny. accomplished is still unknown. Like the other catalytic RNAs, the hairpin ribozyme reaction requires a divalent cation. However, it appears that, unlike the hammerhead ribozyme, the hydrated cation (usually $Mg^{2+}$) is not directly coordinated to the phosphate; moreover, the essential role of divalent cations has recently been called into question.

Since the 'substrate' portion of the cis-cleaving hairpin ribozyme is discontinuous with the rest of the molecule, it is obvious where to separate the two domains. The substrate should contain the sequence RYN*GUC, where R is a purine, Y is a pyrimidine and N is any nucleotide. Cleavage occurs at the position indicated by a*. Stem II should be four base pairs, but stem you can do significantly extend (e.g.,), as can stem IV. Substrate specificity is changed by altering the nonconserved residues within the base-paired region. The hairpin ribozyme is used to target HIV-1 RNA in cell culture, and it is currently approved for clinical trials. However, its effective use as a trans-cleaving ribozyme in vivo is still rather limited; the reasons for this are unclear.

## VS RNA Ribozyme

The mitochondria of certain strains of Neurospora contain the Varkud plasmid (a retroplasmid), which encodes a reverse transcriptase, and a small, unrelated, RNA (VS RNA). The VS RNA is transcribed from circular or multimeric VS plasmid DNA by a mitochondrial RNA polymerase, and the resulting transcripts are subsequently site-specifically cleaved and ligated to form circular, 881 nucleotides long, RNA monomers. These monomers are then reverse transcribed and made double stranded to form the mature VS plasmid.

In vitro transcribed VS RNA precursors are cleaved and ligated by the RNA itself and this is presumed to occur in vivo as well. Of all the self-cleaving RNAs, the catalytic properties of VS RNA are the most poorly understood. At 154 nucleotides long it is also the largest. The minimal sequence that retains catalytic activity contains one nucleotide 5′ and 153 nucleotides 3′ to the cleavage site. However, this structure can be reduced to 121–126 nucleotides by making internal deletions within the helices. An RNA secondary structure is proposed, but except for a tertiary interaction between loop I and loop V, little is known about its overall conformation.

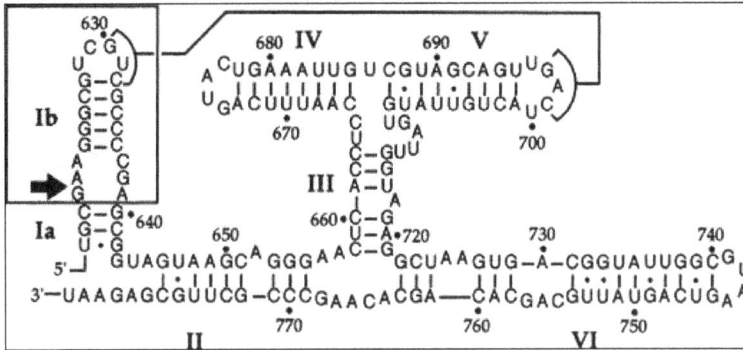

Secondary structure of the VS RNA ribozyme. The arrow shows the cleavage site and numbering is that of the full-length VS RNA. The trans-cleaving form consists of nucleotides 640–881 and the substrate is shown boxed.

The catalytic domain of VS RNA is converted into a trans-cleaving ribozyme by using a 144-nucleotide fragment of the VS RNA from 640 to 881 (VS RNA numbering;). The minimal substrate consists of one nucleotide 5′ and 19 nucleotides 3′ to the cleavage site, and it forms a short stem-loop structure. As with RNase P, the ribozyme seems to recognize the structure of the substrate largely as a helical domain. The minimal sequence requirement 5′ to the cleavage site is a characteristic shared only with the HDV ribozyme and it could make this ribozyme suitable for 3′ end trimming of RNAs expressed in vitro or in vivo. However, the uncertainty in the substrate requirements and the lackadaisical activity in trans have limited its application, although recent experiments have improved its activity.

## HDV Ribozyme

The hepatitis delta virus (HDV) is a viroid-like satellite virus of the hepatitis B virus (HBV), and it is the sole example of such a virus in mammalian systems. It is widespread and can cause severe fulminant hepatitis in infected patients. It is about 1700 nucleotides long, and it encodes a single protein that is expressed in two forms due to an RNA editing event. Both the genomic, infectious strand, and the antigenomic strand have self-cleaving domains. Despite previous pronouncements, no biologically relevant, RNA-catalyzed, ligation reaction has been observed in vitro, although the integrity of the RNA catalytic domains is clearly essential for both the cleavage and ligation reaction in vivo. A possible mechanism for the biological control of these reactions, to prevent inappropriate cleavage or ligation, has been proposed.

The minimal domain containing self-cleaving activity has one nucleotide 5' and 84 nucleotides 3' to the cleavage site for both domains. Despite the sequence differences, both sequences fold into similar secondary structures, of which the pseudoknotted structure shown in figure is now the most widely accepted. It consists of four stem regions; three of these stems (I, II and IV) are largely structural elements, while the specific sequences in hairpin III and in the junctions I/IV and IV/II are more important. The catalytic domains of HDV are known for their ability to retain cis-cleaving activity at high temperatures and in the presence of denaturants. The tertiary structures have been computer modeled for the genomic and antigenomic forms of the pseudoknot model and for the antigenomic axehead variant.

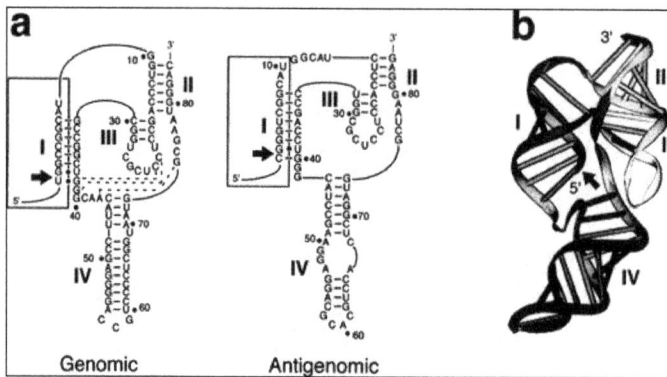

Characteristics of the HDV ribozyme. a: The genomic and antigenomic ribozymes from HDV. Numbering of the nucleotides is relative to the cleavage sites, indicated by arrows. Helical domains are separated by lines to facilitate the presentation. The length of stem I is critical, although there is no evidence that the −1 base pair is needed. The typical trans-cleaving forms are separated in junction I/II, where the substrates are shown boxed. The identities of the nonbase-paired residues in the substrate (boxed) are not important for trans-cleaving activity. Additional base-pair interactions, that were recently derived for the genomic ribozyme, are shown as dashed lines. b: Computer-generated three-dimensional model of the genomic catalytic domain. This figure is modified from. Recently, the crystal structure for the genomic HDV ribozyme was obtained. This structure is similar to that shown here except that there is an additional pseudoknot interaction between C21 and C22 with G38 and G39. Moreover, G10 is stacked on helix II and A43 and G74 stack on hairpin IV.

Recently, a 2.3-Å crystal structure has been solved for the genomic HDV ribozyme. This was accomplished by replacing hairpin IV with a small hairpin structure that binds tightly to the protein U1A, a spliceosomal protein. By co-crystallizing the RNA with the protein, the authors were better able to obtain highly structured crystals that diffracted to a high resolution. It also greatly facilitated heavy metal substitution that is necessary for obtaining crystal phasing. This structure is very similar to the computer-predicted model, but it revealed some unexpected results. There is an additional pseudoknot structure derived from base pairs between C21 and C22, in loop III with

G39 and G38, at the base of helix I. These specific interactions were not previously predicted, although the importance of the nucleotides were correctly derived In addition, A43 and G74 stack on the end of helix IV to form noncanonical base pairs and G10 forms an extension to helix II as previously predicted. These interactions create a structure where helix IV is rotated relative to the computer model and hairpin III is more compressed. Nevertheless, the two structures are otherwise very similar. The crystal structure provides an organized, almost protein-like, crevice for the active site. Unfortunately, it reveals little about the reaction mechanism or the role of the $Mg^{2+}$ ions.

As with the other ribozymes, the cis-cleaving activity of the HDV ribozymes can be converted into a trans-cleaving activity. The most common form is indicated in figure. However, other permutations are possible, including a large substrate consisting of sequences −5 to ~60 and a small ribozyme consisting of sequences from ~60 to +84. Since most of the conserved elements are contained within the substrate, the practical utility of this latter form is somewhat limited. A completely closed-circular variant of the trans-cleaving ribozyme shown in figure also has been generated; a short loop was used to close the end of stem II. The absence of free ends makes this ribozyme particularly resistant to the exonucleases found in serum and the cellular environment. It is possible to change the substrate binding sequence to target other RNAs. In theory, most substrate sequences are possible, although a guanosine at the −1 position, relative to the cleavage site, is inhibitory and a purine-pyrimidine base pair at position +1 is preferred. However, in practice many of the changes in the substrate-binding sequence have unpredictable effects.

## Enzyme Action

An enzyme molecule (E) and the substrate molecule or molecules (S) collide and react to form an intermediate compound called the enzyme-substrate (E–S) complex. (This step is reversible because the complex can break apart into the original substrate or substrates and the free enzyme.) Once the E–S complex forms, the enzyme is able to catalyze the formation of product (P), which is then released from the enzyme surface:

$$S + E \rightarrow E\text{–}S$$

$$E\text{–}S \rightarrow P + E$$

Hydrogen bonding and other electrostatic interactions hold the enzyme and substrate together in the complex. The structural features or functional groups on the enzyme that participate in these interactions are located in a cleft or pocket on the enzyme surface. This pocket, where the enzyme combines with the substrate and transforms the substrate to product is called the active site of the enzyme. It possesses a unique conformation (including correctly positioned bonding groups) that is complementary to the structure of the substrate, so that the enzyme and substrate molecules fit together in much the same manner as a key fits into a tumbler lock. In fact, an early model describing the formation of the enzyme-substrate complex was called the lock-and-key model. This model portrayed the enzyme as conformationally rigid and able to bond only to substrates that exactly fit the active site.

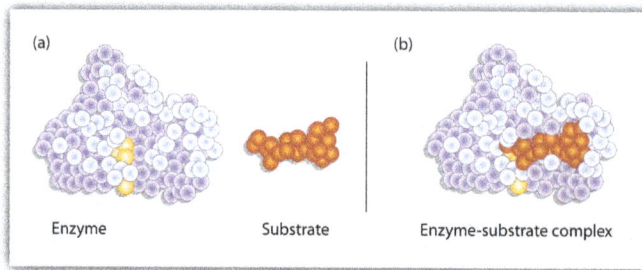

Substrate Binding to the Active Site of an Enzyme. The enzyme dihydrofolate reductase is shown with one of its substrates: NADP+ (a) unbound and (b) bound. The NADP+ (shown in red) binds to a pocket that is complementary to it in shape and ionic properties.

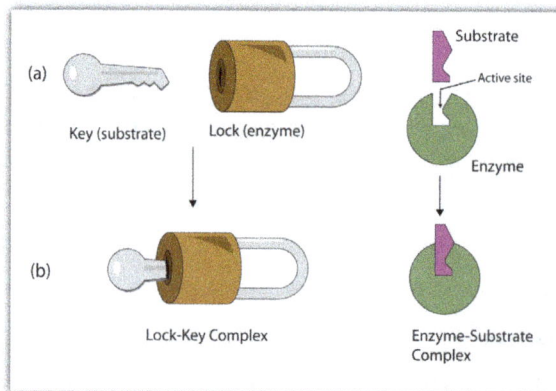

The Lock-and-Key Model of Enzyme Action (a) Because the substrate and the active site of the enzyme have complementary structures and bonding groups, they fit together as a key fits a lock. (b) The catalytic reaction occurs while the two are bonded together in the enzyme-substrate complex.

Working out the precise three-dimensional structures of numerous enzymes has enabled chemists to refine the original lock-and-key model of enzyme actions. They discovered that the binding of a substrate often leads to a large conformational change in the enzyme, as well as to changes in the structure of the substrate or substrates. The current theory, known as the induced-fit model, says that enzymes can undergo a change in conformation when they bind substrate molecules, and the active site has a shape complementary to that of the substrate only after the substrate is bound, as shown for hexokinase in Figure "The Induced-Fit Model of Enzyme Action". After catalysis, the enzyme resumes its original structure.

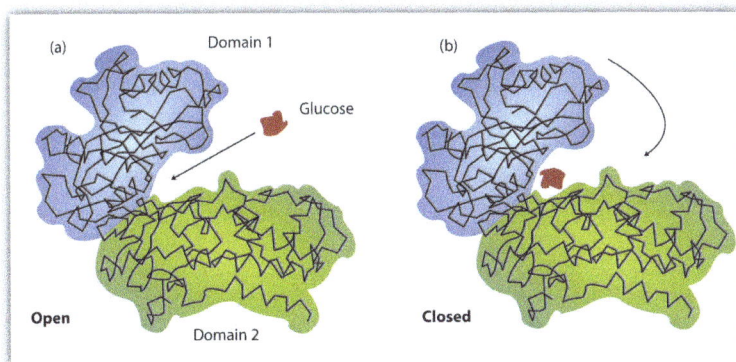

The Induced-Fit Model of Enzyme Action (a) The enzyme hexokinase without its substrate (glucose, shown in red) is bound to the active site. (b) The enzyme conformation changes dramatically when the substrate binds to it, resulting in additional interactions between hexokinase and glucose.

The structural changes that occur when an enzyme and a substrate join together bring specific parts of a substrate into alignment with specific parts of the enzyme's active site. Amino acid side chains in or near the binding site can then act as acid or base catalysts, provide binding sites for the transfer of functional groups from one substrate to another or aid in the rearrangement of a substrate. The participating amino acids, which are usually widely separated in the primary sequence of the protein, are brought close together in the active site as a result of the folding and bending of the polypeptide chain or chains when the protein acquires its tertiary and quaternary structure. Binding to enzymes brings reactants close to each other and aligns them properly, which has the same effect as increasing the concentration of the reacting compounds.

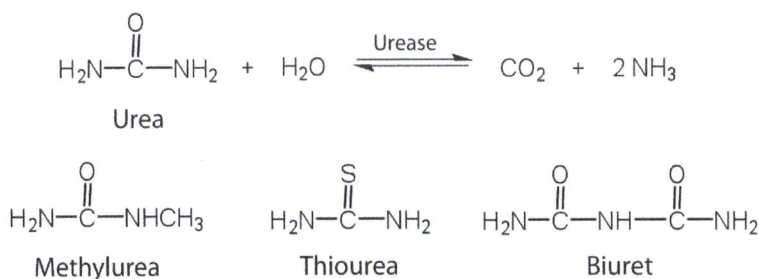

$$\underset{\text{Urea}}{H_2N-\overset{\displaystyle O}{\overset{\|}{C}}-NH_2} \;+\; H_2O \;\underset{}{\overset{\text{Urease}}{\rightleftharpoons}}\; CO_2 \;+\; 2\,NH_3$$

$$\underset{\text{Methylurea}}{H_2N-\overset{\displaystyle O}{\overset{\|}{C}}-NHCH_3} \qquad \underset{\text{Thiourea}}{H_2N-\overset{\displaystyle S}{\overset{\|}{C}}-NH_2} \qquad \underset{\text{Biuret}}{H_2N-\overset{\displaystyle O}{\overset{\|}{C}}-NH-\overset{\displaystyle O}{\overset{\|}{C}}-NH_2}$$

One characteristic that distinguishes an enzyme from all other types of catalysts is its substrate specificity. An inorganic acid such as sulfuric acid can be used to increase the reaction rates of many different reactions, such as the hydrolysis of disaccharides, polysaccharides, lipids, and proteins, with complete impartiality. In contrast, enzymes are much more specific. Some enzymes act on a single substrate, while other enzymes act on any of a group of related molecules containing a similar functional group or chemical bond. Some enzymes even distinguish between D- and L-stereoisomers, binding one stereoisomer but not the other. Urease, for example, is an enzyme that catalyzes the hydrolysis of a single substrate—urea—but not the closely related compounds methyl urea, thiourea, or biuret. The enzyme carboxypeptidase, on the other hand, is far less specific. It catalyzes the removal of nearly any amino acid from the carboxyl end of any peptide or protein.

Enzyme specificity results from the uniqueness of the active site in each different enzyme because of the identity, charge, and spatial orientation of the functional groups located there. It regulates cell chemistry so that the proper reactions occur in the proper place at the proper time. Clearly, it is crucial to the proper functioning of the living cell.

## References

- What-are-enzymes-definition-lesson-quiz: study.com, Retrieved 2 June, 2019

- Enzyme, science: britannica.com, Retrieved 18 February, 2019

- Datta SP, Smith GH, Campbell PN (2000). Oxford Dictionary of Biochemistry and Molecular Biology (Rev. ed.). Oxford: Oxford Univ. Press. ISBN 978-0-19-850673-7

- Enzymes: chemistryexplained.com, Retrieved 9 August, 2019

- Oxidoreductase-Introduction: creative-enzymes.com, Retrieved 27 April, 2019

- Hydrolase-Introduction: creative-enzymes.com, Retrieved 7 January, 2019

- Enzyme nomenclature, 1978 recommendations of the Nomenclature Committee of the International Union of Biochemistry on the nomenclature and classification of enzymes. New York: Academic Press. 1979. ISBN 9780323144605

- Lyase-Introduction: creative-enzymes.com, Retrieved 17 May, 2019

- Hexokinase: chemistryexplained.com, Retrieved 19 March, 2019

- Maltase, enzymes-introduction: worldofenzymes.info, Retrieved 9 July, 2019

- Järvelä I, Torniainen S, Kolho KL (2009). "Molecular genetics of human lactase deficiencies". Annals of Medicine. 41(8): 568–75. doi:10.1080/07853890903121033. PMID 19639477

- Frames: biology.kenyon.edu, Retrieved 29 March, 2019

- What-are-Ribozymes, life-sciences: news-medical.net, Retrieved 1 June, 2019

- Enzyme-action, text-the-basics-of-general-organic-and-biological-chemistry: saylordotorg.github.io, Retrieved 21 April, 2019

- Ranji, A.; Boris-Lawrie, K. (2010). "RNA helicases: Emerging roles in viral replication and the host innate response". RNA Biology. 7 (6): 775–787. doi:10.4161/rna.7.6.14249. PMC 3073335. PMID 21173576

# Enzyme Reaction Kinetics and Enzyme Preparation

**2**

The study of chemical reactions which are catalyzed by enzymes is known as enzyme reaction kinetics. These reactions can be affected by molecules known as enzyme inhibitors and activators. The enzyme inhibitors reduce or abolish the activity of enzymes while enzyme activators increase the catalytic rate of enzymes. This chapter discusses in detail the theories and concepts related to enzyme reaction kinetics and enzyme preparation.

## Enzyme Kinetics

Enzyme kinetics involves the measurement of the rate at which chemical reactions that are catalyzed by enzymes occur. Knowledge about the kinetics of an enzyme can reveal useful information about its catalytic mechanism, role in metabolism, factors that impact its activity, and mechanisms of inhibition.

### Rate of Reaction

Enzymes are thought to form a complex with the substrates to catalyze the reaction. This process can be illustrated with the simplified equation, where e is the enzyme, S is the substrate, and P is the product:

$$E + S \Leftrightarrow ES \Rightarrow P + E$$

The first step of the equation, which is reversible, has the reaction rate constant of k+1 to produce the enzyme substrate complex and $k_{-1}$ for the reverse reaction. The reaction rate constant for the second step of the equation, which is not reversible, is $k_{+2}$.

The rate of reaction (v), which is the rate at which the product is formed, is defined by the following equation:

$$v = d[P]/dt = k_{+2}[ES]$$

The square brackets in the above equation represent the molar concentration of the substance specified within, so [P] refers to the molar concentration of the product and [ES] the molar concentration of the enzyme substrate.

# $V_{max}$ and $K_m$

Leonor Michaelis and Maud Menten introduced a mathematical illustration to describe the action of enzymes with two constants, $V_{max}$ and $K_m$.

The maximal velocity ($V_{max}$) refers to the point at which the increase the concentration of the substrate does not increase the rate of a reaction catalyzed by an enzyme. This occurs because the substrate molecules saturate the active sites of the enzyme and are not able to form more complexes with the enzyme. This value is given as a rate (mol/s), which is the maximum velocity of the reaction when the enzyme is saturated.

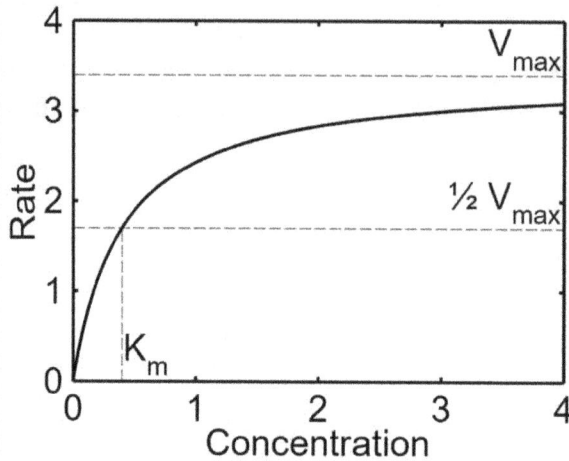

Michaelis–Menten saturation curve of an enzyme reaction. Parameter values used are Vmax=3.4 and Km=1.7.

The Michaelis constant ($K_m$) is the concentration of the substrate when half of the active binding sites of an enzyme are occupied by the substrate. The constant helps to depict the affinity of the enzyme for their substrate. This value is given as the concentration of the substrate (mM) at half of $V_{max}$. An enzyme with a high $K_m$ has a low affinity for the substrate, and a high concentration of the substrate is needed in order for the enzyme to become saturated. Conversely, an enzyme with a high $K_m$ has a high affinity for the substrate and the enzyme may become saturated even with a low amount of substrate.

## Applications of Enzyme Kinetics

There are many practical uses of enzyme kinetics. For example, the kinetic constants can help explain how enzymes work and assist in the prediction of the behavior of enzymes in living organisms.

$V_{max}$ and $K_m$ both play a key role in understanding the metabolism of the human body. Knowledge of the enzyme kinetic constants allows us to gain a better understanding of the enzymes and processes that take place in human metabolism.

# Enzyme Inhibitors

Enzyme inhibitors are molecules that interact in some way with the enzyme to prevent it from working in the normal manner.

Enzymes are different chemical compounds that are combined into a group because of their only feature—they can suppress enzyme activity. The suppression of the activity is the result of the binding of inhibitor to the enzyme molecule that arrests catalytic reaction. Because enzymes catalyze most part of chemical reactions in living organisms, the enzyme inhibitors play an important role in the development of different sciences (biochemistry, physiology, pharmacy, agriculture, ecology) as well as the technologies (production of pharmaceutical drugs, insecticides, pesticides, chemical weapons, etc.).

Many pharmacological drugs are enzyme inhibitors. The group of well-known pharmaceutical agents with name nonsteroidal antiinflammatory drugs (NSAIDs) includes inhibitors of enzyme cyclooxygenase that catalyzes a first step of synthesis of biologically active compounds prostaglandins that are responsible for the development of pain, inflammation, fever, contraction of smooth muscle, formation of blood clots, and others.

All inhibitors may be combined in different groups in accordance with their chemical structure: ions of metals ($Hg^+$, $Fe^{2+}$, $Cu^+$, $Pb^{2+}$), organic compounds (e.g., N-ethylmaleimide, diisopropyl phosphofluoridate, oligomycin), and large bioorganic molecules, (peptides, proteins, etc). However, this classification does not reflect mechanism of their interaction with enzyme.

In accordance with the mode of action, enzyme inhibitors may be divided into two different groups (reversible and irreversible inhibitors). Reversible inhibitors, in turn, may be combined in four groups in accordance with kinetic behavior (competitive, uncompetitive, noncompetitive, and mixed inhibitors).

The mechanism of action of enzyme inhibitors includes a step of enzyme-inhibitor complex formation (EI complex) that has no (or low) enzyme activity. An irreversible inhibitor dissociates from this complex very slow because it is tightly bound to the enzyme. Mainly this mode of inhibition is connected with the formation of covalent bond or hydrophobic interaction between enzyme and inhibitor. Irreversible inhibitors usually react with the enzyme and change it chemically. These inhibitors often contain reactive functional groups that modify amino acid residues of enzyme that are essential for its activity. They also can provide inhibition affecting the enzyme conformation. An example of irreversible inhibitor is N-ethylmaleimide that covalently interacts with SH-group of cysteine residues of enzyme molecules, like peptidase (insulin-degrading enzyme), 3-phosphoglyceraldehyde dehydrogenase, or hydrophobic compound from group of cardiotonic steroids that at the last bind to Na,K-ATPase using hydrophobic interactions. Another well-known irreversible inhibitor is diisopropyl phosphofluoridate that modifies OH-group of serine residue in active site of such enzymes as chymotrypsin and other serine proteases or acetylcholine esterase in cholinergic synapsis of the nervous system being a potent neurotoxin. Inhibition of this enzyme causes an increase in the acetylcholine neurotransmitter concentration that results in muscular paralysis and death. Inhibitor of cyclooxygenase aspirin (acetyl salicylic acid) covalently modifies OH-group of serine residue located in a close proximity to the active site of cyclooxygenase.

Irreversible inhibition is different from irreversible enzyme inactivation. Irreversible inhibitors are generally specific for one class of enzymes and do not inactivate all proteins. In contrast to denature agents such as urea, detergents do not destroy protein structure but specifically alter the active site of the target enzyme.

Consequently because of tight binding, it is difficult to remove an irreversible inhibitor from the EI complex after its formation. So, we can refer some chemical compound to irreversible enzyme inhibitor, if after the formation of EI complex, the dilution of it with significant amount of water (100–200 excess) does not restore enzyme activity.

Irreversible inhibitors display time-dependent loss of enzyme activity. Interaction of irreversible inhibitor with enzyme is a bimolecular reaction:

$$E + I \xrightarrow{K_I} EI,$$

where E is enzyme, I is inhibitor, EI is complex of enzyme-inhibitor, and ki is a constant of the velocity of this reaction.

However, usually the action of irreversible inhibitors is characterized by the constant of observed pseudo-first order reaction under conditions when concentration of inhibitor is significantly higher than concentration of the enzyme. The value of pseudo-first order rate of inhibition may be measured by plotting of the ln of enzyme activity (in % relatively enzyme activity in the absence of inhibitor) vs. time. Tangent of slope angle of straight line obtained by this way will be equal to value of constant of pseudo-first order inhibition. The value of rate constant of bimolecular reaction for irreversible inhibition may be then calculated by dividing the obtained value of constant of pseudo-first order reaction per inhibitor concentration.

Reversible inhibitor binds to the enzyme reversibly. It means that there is equilibrium between the formation and dissociation of EI complex:

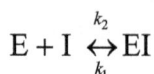

$$E + I \underset{k_1}{\overset{k_2}{\leftrightarrow}} EI$$

where $k_1$ is a constant of the velocity of direct reaction and $k_2$ is a constant of the velocity of reverse reaction. The effect of reversible inhibitors is characterized by the constant of dissociation of EI complex that is equal to [E] [I]/[EI] or $k_1/k_2$.

Usually reversible inhibitor binds to the enzymes using non-covalent interactions such as hydrogen or ionic bonds. Different types of reversible inhibition are produced depending on whether these inhibitors bind to the enzyme, the enzyme-substrate complex, or both.

One type of reversible inhibition is called competitive inhibition. In this case, there are two types of complexes: enzyme inhibitor (EI) and enzyme substrate (ES); complex EI has no enzyme activity. The substrate and inhibitor cannot bind to the enzyme at the same time. This inhibition may be reversed by the increase of substrate concentration. However, the value of maximal velocity ($V_{max}$) remains constant. The value of apparent Km will increase; however, the value of maximal velocity ($V_{max}$) remains constant. It can be competitive inhibition not only in relation to substrate but also to cofactors, as well as to activators.

Usually reversible inhibitor binds to the enzymes using non-covalent interactions such as hydrogen or ionic bonds. Different types of reversible inhibition are produced depending on whether these inhibitors bind to the enzyme, the enzyme-substrate complex, or both.

One type of reversible inhibition is called competitive inhibition. In this case, there are two types of complexes: enzyme inhibitor (EI) and enzyme substrate (ES); complex EI has no enzyme activity. The substrate and inhibitor cannot bind to the enzyme at the same time. This inhibition may be reversed by the increase of substrate concentration. However, the value of maximal velocity (Vmax) remains constant. The value of apparent Km will increase; however, the value of maximal velocity (Vmax) remains constant. It can be competitive inhibition not only in relation to substrate but also to cofactors, as well as to activators.

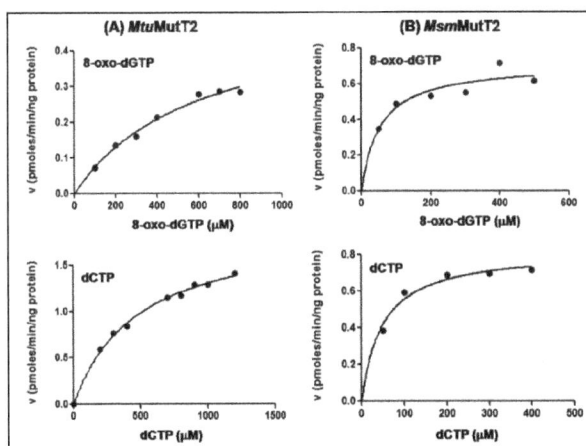

Kinetic test for reversible inhibitor classification.

Another type of reversible inhibition is called uncompetitive inhibition. In this case, the inhibitor binds only to the substrate-enzyme complex; it does not interfere with the binding of substrate with active site but prevents the dissociation of complex enzyme substrate: it resulted in the dependence of the inhibition only upon inhibitor concentration and its Ki value. This type of inhibition results in $V_{max}$ decrease and Km decrease.

The third type of inhibition is noncompetitive. This type of inhibition results in the inability of complex enzyme (E) inhbitor (I) substrate (EIS) to dissociate giving a product of reaction. In this case, inhibitor binds to E or to ES complex. The binding of the inhibitor to the enzyme reduces its activity but does not affect the binding of substrate. As a result, the extent of the inhibition depends only upon the concentration of the inhibitor. In this case, Vmax will decrease, but Km will remain the same.

In some cases, Mixed Inhibition can be seen, when the inhibitor can bind to the enzyme at the same time as to enzyme-substrate complex. However, the binding of the inhibitor effects on the binding of the substrate and vice versa. This type of inhibition can be reduced, but not overcome by the increase of substrate concentrations. Although it is possible for mixed-type inhibitors to bind in the active site, this inhibition generally results from an allosteric effect of inhibitor. An inhibitor of this kind will decrease Vmax, but it will increase Km.

Special case of enzyme inhibition is inhibition by the excess of substrate or by the product. This inhibition may follow the competitive, uncompetitive, or mixed patterns. Inhibition of enzyme by its substrate occurs when a dead-end enzyme-substrate complex forms. Often in the case of substrate inhibition, a molecule of substrate binds to active site in two points (e.g., by the "head" and by the "tail" of molecule). At high concentrations, two substrate molecules bind in active site the following manner: one substrate molecule binds using the "head" and another molecule using the "tail." This binding is nonproductive and substrate cannot be converted to the product. An example of such inhibition is inhibition of acetyl cholinesterase by the excess of acetylcholine.

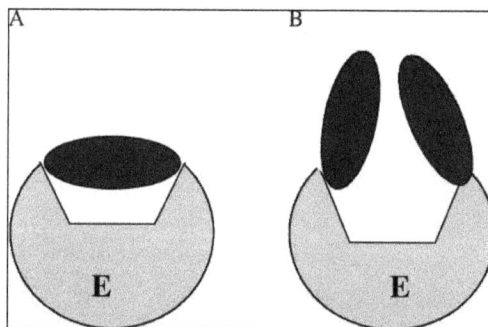

Enzyme inhibition by substrate. Productive binding of one substrate molecule with two points of enzyme active site (A) and unproductive binding of two substrate molecules with the same site (B)

Competitive inhibitors mainly interact with enzyme active site preventing binding of real substrate. Classical example of competitive inhibition is inhibition of fumarate hydratase by maleate that is a substrate analog. Enzyme is highly stereospecific; it catalyzes the hydration of the trans-double bound of fumarate but not maleate (cis-isomer of fumarate). Maleate binds to active site with high affinity preventing the binding of fumarate. Despite the binding maleate to active site, it cannot be converted into the product of reaction. However, maleate occupies active site making it inaccessible for real substrate and providing by this way the inhibition.

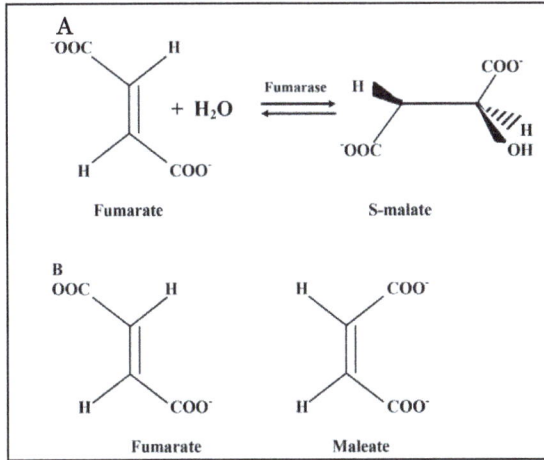

Example of enzyme competitive inhibitors. A reaction catalyzing by fumarate hydratase (A) and comparison of structure of fumarate (substrate of reaction) and maleate (enzyme competitive inhibitor) (B)

Some reversible inhibitors bind so tightly to the enzyme that they are essentially irreversible. It is known that proteolytic enzymes of the gastrointestinal tract are secreted from the pancreas in an inactive form. Their activation is achieved by restricted trypsin digestion of proenzymes. To stop activation of proteolytic enzymes, the pancreas produces trypsin inhibitor. It is a small protein molecule (it consists of 58 amino acid residues). This inhibitor binds directly to trypsin active site with Kd value that is equal to 0.1 pM. The binding is almost irreversible; complex EI does not dissociate even in solution of 6 M urea. The inhibitor is a very effective analog of trypsin substrates; amino acid residue Lys-15 of inhibitor molecule interacts with aspartic residue located in a pocket of enzyme surface destined for substrate binding, thereby preventing its binding and conversion into the product.

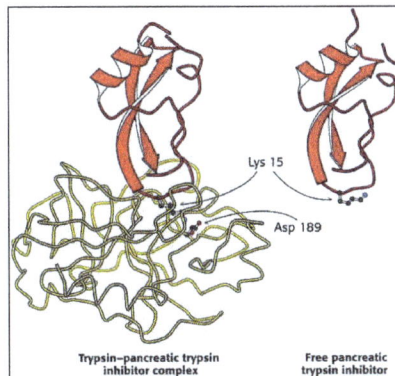

Structure of complex pancreatic trypsin inhibitor—trypsin and free trypsin inhibitor.

Irreversible inhibitors as a tool for study of enzymes: Enzyme active sites labeling by irreversible inhibitors.

To obtain information concerning the mechanism of enzyme reaction, we should determine functional groups that are required for enzyme activity and located in enzyme active site. First approach is to reveal a 3D structure of enzyme with bound substrate using X-ray crystallography. Alternative and additional approach is to use group-specific reagent that simultaneously is irreversible inhibitor of the enzyme. It can covalently bind to reactive groups of enzyme active site that allow to elucidate functional amino acid residues of the site. Modified amino acid residues may be found later after achievement of complete enzyme inhibition, enzyme proteolysis, and identification of labeled peptide(s).

Irreversible inhibitors that can be used with this aim may be divided into two groups: (1) group-specific reagents for reactive chemical groups and (2) substrate analogs with included functional groups that are able to interact with reactive amino acid residues. These compounds can covalently modify amino acids essential for activity of enzyme active site and in such a manner can label them.

One from the most known group-specific reagent that was used to label functional amino acid residue of enzyme active site of protease chymotrypsin was diisopropyl phosphofluoridate. It modified only 1 from 28 serine residues of the enzyme. It means that this serine residue is very reactive. Location of Ser-195 in active site of chymotrypsin was confirmed in investigation carried out later, and the origin of its high reactivity was revealed. Diisopropyl phosphofluoridate was also successfully used for identification of a reactive serine residue in the active site of acetylcholinesterase.

To reveal reactive SH-group in active site of various enzymes, different SH-reagents were used, among them 14C-labeled N-ethylmaleimide, iodoacetate, and iodoacetamide. Using these reagents, cysteines were revealed in the active sites of some dehydrogenase, cysteine protease, and other enzymes.

The second approach is the application of reactive substrate analogs. These compounds are structurally similar to the substrate but include chemically reactive groups, which can covalently bind to some amino acid residues. Substrate analogs are more specific than group-specific reagents. Tosyl-L-phenylalanine chloromethyl ketone, a substrate analog for chymotrypsin that is able to bind covalently with histidine residue and irreversibly inhibit enzyme, makes possible identification of Hys-57 in chymotrypsin active site.

## Natural Enzyme Inhibitors

Many cellular enzyme inhibitors are proteins or peptides that specifically bind to and inhibit target enzymes. Numerous metabolic pathways are controlled by these specific compounds that are synthesized in organisms. Very interesting example of these inhibitors is protein serpins. It is a large family of proteins with similar structures. Most of them are inhibitors of chymotrypsin-like serine protease.

Serine proteases (e.g., mentioned above chymotrypsin) possess a reactive serine residue in active site and have similar mechanisms of catalysis. Cleavage of peptide bond by these proteases is a two-step process. Reactive serine residue of the protease active site that looses $H^+$ and becomes nucleophilic one in the beginning of catalytic act attacks substrate peptide bond. This results in the release of new N-terminal part of protein substrate (first product) and in the formation of a covalent ester bond between the enzyme and the second part of substrate. The second step of catalysis of usual substrates leads to the hydrolysis of ester bond and to the release of the second product (C-terminal part of protein substrate). If serpin is cleaved by a serine protease, it undergoes conformational transition before the hydrolysis of ester bond between enzyme and the second part of substrate (serpin). The change of serpin conformation leads to the "freezing" of intermediate (complex of enzyme with covalently attached second part of serpin is retained for several days). Therefore, serpins are irreversible inhibitors with unusual mechanism of action. They have named "suicide inhibitors," because each serpin molecule can inactivate a single molecule of protease and kills itself during the process of protease inhibition.

Considering enzyme inhibitors we should keep in mind that many living organisms are in the state of "chemical war." Fungi are fighting with bacteria for food using antibiotics. Most immobile organisms like plants and some sea invertebrates use different poisons to defense themselves from being eaten; some vertebrates (like snakes) and invertebrates (e.g., bee and wasps) use poisons not only for defense but also to get food.

Poisons of plants and invertebrates were used as medicine drugs during thousands of years. But only in the twentieth century, it became clear that the poisons contain various enzyme inhibitors as well as the blockers of some other biological molecules (channels, receptors, etc.) For example, bee venom includes melittin, peptide containing 28 amino acids. This peptide can interact with many enzymes suppressing their activities; in particular, it binds with protein calmodulin that are activator of many enzymes. Special studies have shown that melittin structure imitates structure of some proteins (to be exact, some part of protein molecules) that can interact with target enzyme to provide their biological function.

Another example of natural inhibitors is cardiotonic steroids that were found initially in plants (digoxin, digitonin, ouabain) and in the mucus of toads (marinobufagenin, bufotoxin, etc.). These compounds are irreversible inhibitors of Na, K-ATPase that is enzyme transporting $Na^+$ and $K^+$ through the plasma membrane of animals against the electrochemical gradients. In the end of the twentieth century, it was shown that cardiotonic steroids are presented in low concentrations in the blood of mammals including human beings. The increase of concentration of these compounds in the blood may be involved in the development of several cardiovascular and renal diseases including volume-expanded hypertension, chronic renal failure, and congestive heart failure.

## Enzyme Inhibitors as Pharmaceutical Agents

Inhibitors of angiotensin-converting enzyme (ACE). ACE catalyzes a conversion of inactive decapeptide angiotensin I into angiotensin II by the removal of a dipeptide from the C-terminus of angiotensin I. Angiotensin II is a powerful vasoconstrictor. Inhibition of ACE results in the decrease of angiotensin I concentration and in the relaxation of smooth muscles of vessels. Inhibitors of ACE are widely used as drugs for treatment of arterial hypertension.

Proton pump inhibitors (PPIs). Proton pump is an enzyme that is located in the plasma membrane of the parietal cells of stomach mucosa. It is a P-type ATPase that provides proton secretion from parietal cells in gastric cavity against the electrochemical gradient using energy of adenosine triphosphate (ATP) cleavage. PPIs are groups of substituted benzopyridines that in acid medium of stomach are converted into active sulfonamides interacting with cysteine residues of pump. Therefore, PPIs are acid-activated prodrugs that are converted into drugs inside the organisms.

Statins represent a group of compounds that are analogs of mevalonic acid. They are inhibitors of 3-hydroxy-3-methylglutaryl-CoA reductase, an enzyme participating in cholesterol synthesis. Statins are used as drugs preventing or slowing the development of atherosclerosis. Because of the existence of some adverse effects, statins may be recommended for patients that cannot achieve a decrease of cholesterol level in the blood through diet and changes in lifestyle.

Antibiotic penicillin covalently modifies the enzyme transpeptidase, thereby preventing the synthesis of bacterial cell walls and thus killing the bacteria.

Methotrexate is a structural analog of tetrahydrofolate, a coenzyme for the enzyme dihydrofolate reductase, which catalyzes necessarily step in the biosynthesis of purines and pyrimidines. Methotrexate binds to this enzyme approximately 1000-fold more tightly than the substrate and inhibits nucleotide base synthesis. It is used for cancer therapy.

New promising direction of anticancer therapy that is connected with suppression of protein kinases controlling the cellular response to DNA damage is now on the step of development. Selective inhibitors of these enzymes are now being tested in clinical trials in cancer patients.

# Enzyme Activators

## Enzyme Inhibitors

Enzyme activators are chemical compounds that increase a velocity of enzymatic reaction. Their actions are opposite to the effect of enzyme inhibitors. Among activators we can find ions, small organic molecules, as well as peptides, proteins, and lipids.

There are many enzymes that are specifically and directly activated by small inorganic molecules, mainly by cations such as $Ca^{2+}$ which is a the second messenger (among enzymes activated by $Ca^{2+}$, we can find different regulatory enzymes, in particular phospholipases II, protein kinases C, adenylyl cyclases, etc.). These enzymes usually have special site for $Ca^{2+}$ binding; the binding of $Ca^{2+}$ with it results in the change of enzyme conformation that increase enzyme activity.

Cations can bind not only with enzyme but also with the substrate increasing its affinity to the enzyme that activate enzyme. For example, magnesium ions interact with ATP or with other nucleotides that are negatively charged molecules, decreasing their charge that provides effective binding of nucleotides in substrate binding site of various enzymes and increasing their activity.

In some cases, activation of enzymes is due to the elimination of enzyme inhibitors. In total this effect looks as enzyme activation. Some cations including heavy metal cations inhibit definite enzymes. Small organic compounds like ethylene glycol-bis(β-amino-ethyl ether)-N,N,N',N'-tetraacetic acid (EGTA) and ethylenediaminetetraacetic acid (EDTA) that are known as chelating agents bind these inhibitory cations and by this way can eliminate their inhibitory effect.

Special group of activators can produce activation of target enzymes only after the formation of complex with another molecule. This complex, in turn, binds to enzyme and increases the velocity of enzymatic reaction. The most well-known example of such type of activators is Ca-binding protein calmodulin (calcium-modulated protein) that is expressed in all eukaryotic cells. Calmodulin is a small protein containing 148 amino acids (16.7 kDa). Its molecule consists of two symmetrical globular domains each with two Ca-binding motifs (EF-hand) located on N- and C-domains that are jointed by flexible linker. Flexibility of calmodulin molecule and the presence of nonpolar grooves in the middle part of the protein allow it to bind a large variety of proteins. The binding of $Ca^{2+}$ to calmodulin changes its conformation. These, in turn, make complex calmodulin-$Ca^{2+}$ suitable for interaction with target enzymes (calmodulin-dependent protein kinases and phosphatases,Ca-ATPase of plasma membrane, etc.), by this manner increasing their activity. Therefore calmodulin is considered as a participant of calcium signal transduction pathway that provides enforcing and prolongation of the effect of $Ca^{2+}$ as a second messenger.

## Allosteric Enzyme Modulators

Inhibitors and activators (modulators) that bind to enzymes not in the active site but in special center located far enough from it have name allosteric modulators. Their binding to allosteric sites induces the change of enzyme conformation that affects both the structure of active site and enzyme conformational mobility leading to the decrease or to the increase of enzyme activity. Just as enzyme active site is specific in relation to substrate, the allosteric site is specific to its modulator.

Many metabolic pathways are regulated through the action of allosteric modulators. Enzymes in metabolic pathways work sequentially, and in such pathways, a product of one reaction becomes a substrate for the next one. The rate of whole pathway is limited by the rate of the lowest reaction. Allosteric regulators often are a final product of whole metabolic pathway that activates enzymes catalyzing a limiting step of the whole pathway. Enzymes in a metabolic pathway can be inhibited or activated by downstream products. This regulation represents negative and positive feedbacks that slow metabolic pathway when the final product is produced in large amounts or accelerate it when a final product is presented in low concentration. Therefore, allosteric modulators are important participants of such negative and positive feedbacks in metabolic pathways or between them making metabolism self-controlled.

For example, ATP and citrate are inhibitors of phosphofructokinase that is a key enzyme of glycolytic pathway. One product of glycolysis is ATP. Another product is pyruvate that after the conversion into acetyl-CoA is condensed with citrate opening cycle of citrate acids (Krebs cycle). Reactions of this cycle produce reduced nicotinamide adenine dinucleotide reduced (NADH) and flavinadeninidinucleotide reduced (FADH2), oxidation of which is coupled with massive production of ATP in mitochondria. Availability of ATP or citrate inhibits glycolysis preventing glucose oxidation (negative feedback). Inhibition of phosphofructokinase by ATP or by citrate occurs by allosteric manner. Described negative feedback control maintains a steady concentration of ATP in the cell. metabolic pathways are regulated not only through inhibition but also through activation of the key enzymes. Mentioned above phosphofructokinase is activated by adenosine diphosphate (ADP), adenosine monophosphate (AMP), and fructose-2,6-bisphospate that represents positive feedback control.

Enzymes that are regulated by allosteric modulators are usually presented by several interacting subunits (they are called oligomers). A very interesting example of regulation of the activity of oligomeric enzymes is c-AMP-dependent protein kinase that is an important regulatory enzyme participating in the phosphorylation of serine and threonine residues of target proteins changing by this way their activity. This enzyme consists of four subunits; two of them are catalytic and two are regulatory. Cyclic AMP (c-AMP) is allosteric activator of this enzyme. Catalytic subunit being bound to the regulatory one is inactive. Binding of two c-AMP molecules to allosteric sites of each regulatory subunit induces their conformation transition that results in dissociation of the tetrameric complex and in activation of catalytic subunits. Decrease of c-AMP concentration leads to its dissociation from the allosteric site and to association of regulatory and catalytic subunits with subsequent inactivation of catalytic subunits. By this way, c-AMP activity depends upon the c-AMP concentration in the cell.

## Enzyme Assay

Enzyme assays are laboratory techniques for determining enzymatic activity. They are vital for the study of enzyme kinetics and enzyme inhibition.

Enzyme activity: Enzyme activity is a measure of the quantity of active enzyme present and is thus dependent on conditions, which should be specified. The SI unit is the katal, 1 katal = 1 mol s-1, but this is an excessively large unit. A more practical and commonly-used value is 1 enzyme unit (U) = 1 µmol min-1. 1 U corresponds to 16.67 nanokatals.

Specific activity: The specific activity of an enzyme is another common unit. This is the activity of an enzyme per milligram of total protein (expressed in µmol min-1mg-1). Specific activity gives a measurement of the activity of the enzyme. It is the amount of product formed by an enzyme in a given amount of time under given conditions per milligram of total protein. Specific activity is equal to the rate of reaction multiplied by the volume of reaction divided by the mass of total protein.

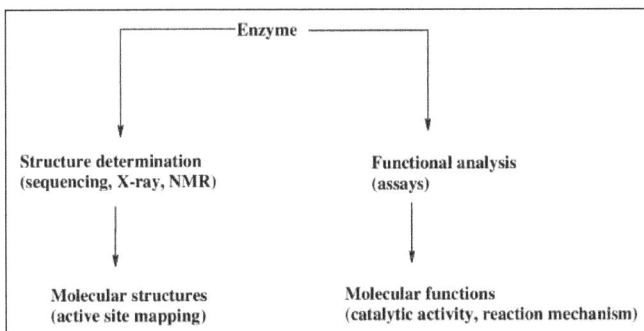

Enzyme assays as tools for functional analysis.

## Enzyme Assay as a Tool for Selection of Active Enzymes

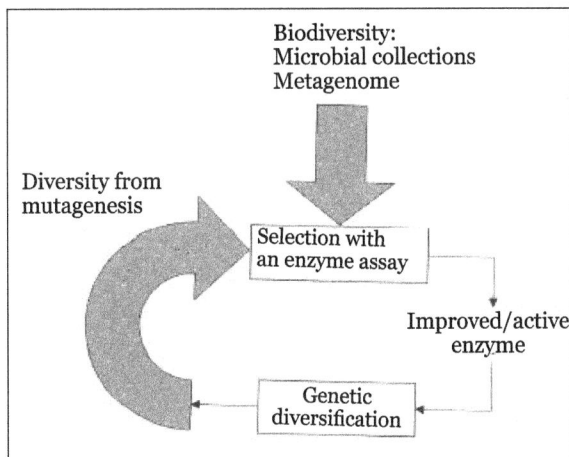

1. Development of an assay

A useful enzyme assay must meet four criteria:

a) Absolute specificity.

b) High sensitivity.

c)  High precision and accuracy.

d)  Convenience.

## 2. High sensitivity

e.g. for purification, specific activities of most enzymes are very low. Therefore, the assay must be highly sensitive and reproducible.

## 3. High precision

The accuracy and precision of an enzyme assay usually depend on the underlying chemical basis of techniques that are used.

For example, if an assay is carried out in buffer of the wrong pH, the observed rates will not accurately reflect the rate of enzymatically produced products

## Six Major Characteristics of a Protein Solution

Six major characteristics of a protein solution warrant consideration

1.  pH.

2.  Degree of oxidation.

3.  Heavy metal contamination.

4.  Medium polarity.

5.  Protease contamination.

6.  Temperature.

## pH

pH values yielding the highest reaction rates are not always those at which the enzyme is most stable. It is advisable to determine the pH optima for enzyme assay and stability separately.

## For Protein Purifications

Buffer must have an appropriate pKa and not adversely affect the protein(s) of interest. Buffer capacity may be higher for tissues with large vacuoles such as plants and fungi.

## Degree of Oxidation

Most proteins contain free SH groups. One or more of these groups may participate in substrate binding and therefore are quite reactive.

Upon oxidation, SH turn form intra- or inter-molecular S-S bonds, which usually result in loss of enzyme activity.

A wide variety of compounds are available to prevent disulfide bond formation: 2-mercaptoethanol, cysteine, reduced glutathione, and thioglycolate. These compounds are added to protein solutions at concentration ranging from $10^{-4}$ to $5 \times 10^{-3}$ M (excess because equilibrium are near unity).

Dithiothreitol is advantageous (lower amounts needed) because of formation of stable six-ring.

Antioxidants against quinones (e.g. protein isolation from plants) by polyvinylpyrrolidone.

## Heavy Metal Contamination

SH groups may react with heavy metal ions such as Pb, Fe, Cu stemming from buffers, ion exchange resins or even the water in which solutions are prepared.

If trace amounts of heavy metals continue to be a problem, EDTA (ethylenediaminetetraacetic acid) may be included in the buffer solutions at a concentration of 1 to $3 \times 10^{-4}$ M. The compounds chelates most, if not all, deleterious metal ions.

## Protease or Nuclease Contamination

During cell breakage, proteases and nucleases are liberated.

PMSF (phenylmethylsulfonyl fluoride): a commonly used protease inhibitor.

## Temperature

Not all proteins are most stable at 0 °C, e.g. Pyruvate carboxylase is cold sensitive and may be stabilized only at 25 °C.

Freezing and thawing of some protein solutions is quite harmful. If this is observed, addition of glycerol or small amounts of dimethyl sulfoxide to the preparation before freezing may be of help.

Storage conditions must be determined by trial and error for each protein.

## More on Keeping Proteins for Enzyme Assays

Proteins requiring a more hydrophobic environment may be successfully maintained in solutions whose polarity has been decreased using sucrose, glycerol, and in more drastic cases, dimethyl sulfoxide or dimethylformamide. Appropriate concentrations must usually be determined by trial and error but concentrations of 1 to 10% (v/v) are not uncommon.

few proteins, on the other hand, require a polar medium with high ionic strength to maintain full activity. For these infrequent occasions, KCl, NaCl, $NH_4Cl$, or $(NH_4)_2SO_4$ may be used to raise the ionic strength of the solution.

Protein purification for testing novel enzymes: series of isolation and concentration procedures.

Major techniques for the isolation and concentration of proteins : differential solubility, ion exchange chromatography, absorption chromatography, molecular sieve techniques, affinity chromatography, electrophoresis.

## Coupled Enzyme Assays

• If neither the substrates nor products of an enzymecatalyzed reaction absorb light at an appropriate wavelength,the enzyme can be assayed by linking to another enzyme-catalyzed reaction that does involve a change in absorbance.

• The second enzyme must be in excess,so that the rate-limiting step in the linked assay is the action of the first enzyme.

## Outline of an HTS-enzyme Assays

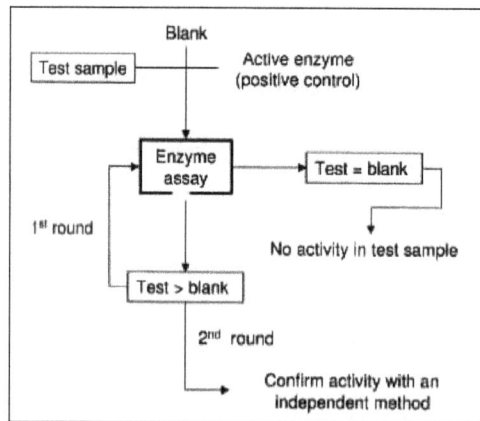

p-Nitrophenyl derivatives for chromogenic and fluorogenic assay for hydrolase activity.

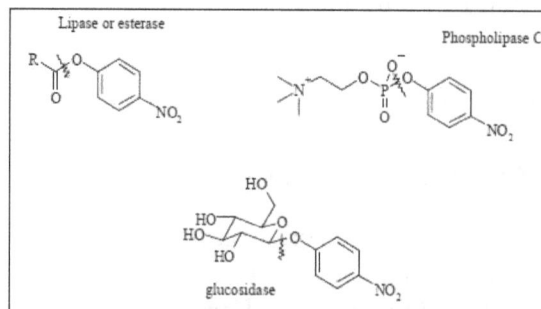

Typical p-nitrophenyl derivatives for chromogenic assay of hydrolases. After hydrolysis each substrate releases p-nitrophenol which can be detected by UV-spectrophotometer.

# Enzyme Preparation

## Enzyme Extraction

Enzymatic extraction uses enzymes to degrade the cell walls of algae with water acting as the solvent, this makes fractionation of the oil much easier. The costs of this extraction process are estimated to be much greater than hexane extraction.

## Enzyme Purification

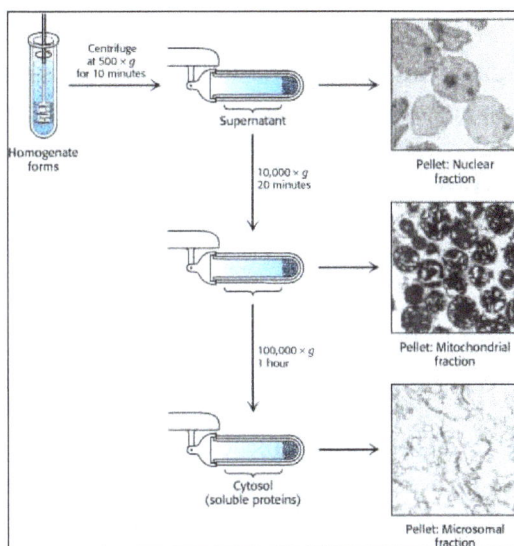

Figure: Differential Centrifugation.

Purification and separation of enzymes are generally based on solubility, size, polarity, and binding affinity. The production scale, timeline, and properties of the enzymes should all be considered when choosing the proper separation method.

1. Solubility based separation

The principle of the type of separation is that enzyme solubility changes drastically when the pH, ionic strength, or dielectric constant changes. For example, most proteins are less soluble at high salt concentrations, an effect called salting out. The salt concentration at which a protein precipitates differs from one protein to another. Hence, salting out can be used to fractionate proteins. Salting out is also useful for concentrating dilute solutions of proteins, including active fractions obtained from other purification steps. Addition of water-miscible organic solvents such as ethanol or acetone will change the dielectric constant of the solvent and therefore precipitate the desired enzyme. Neutral water-soluble polymers can also be used for the same purpose instead of organic solvents. However, the risks of losing enzyme activity during precipitation and further separation of the added salt or polymer need to be considered.

## 2. Size or mass based method

Because enzymes are relatively large molecules, separation based on the size or mass of molecules favors purification of enzymes, especially the ones with high molecular weight. Dialysis is a commonly used method, where semipermeable membranes are used to remove salts, small organic molecules, and peptides. The process usually needs a large volume of dialysate, the fluid outside the dialysis dag, and a period of hours or days to reach the equilibrium. Countercurrent dialysis cartages can also be used, in which the solution to be dialyzed flow in one direction, and the dialysate in the opposite direction outside of the membrane. Similarly, ultrafiltration membranes, which are made from cellulose acetate or other porous materials, can be used to purify and concentrate an enzyme larger than certain molecular weight. The molecular weight is called the molecular weight cutoff and is available in a large range from different membranes. The ultrafiltration process is usually carried out in a cartridge loaded with the enzyme to be purified. Centrifugal force or vacuum is applied to accelerate the process. Both dialysis and ultrafiltration are quick but somewhat vague on distinguishing the molecular weight, whereas size exclusion chromatography gives fine fractionation from the raw mixture, allowing separation of the desired enzyme from not only small molecules but also other enzymes and proteins. Size exclusion chromatography, also known as gel-filtration chromatography, relies on polymer beads with defined pore sizes that let particles smaller than a certain size into the bead, thus retarding their egress from a column. In general, the smaller the molecule, the slower it comes out of the column. Size exclusion resins are relatively "stiff" and can be used in high pressure columns at higher flow rates, which shortens the separation time. Other factors including the pore size, protein shape, column volumes, and ionic strength of the eluent could also change the result of purification.

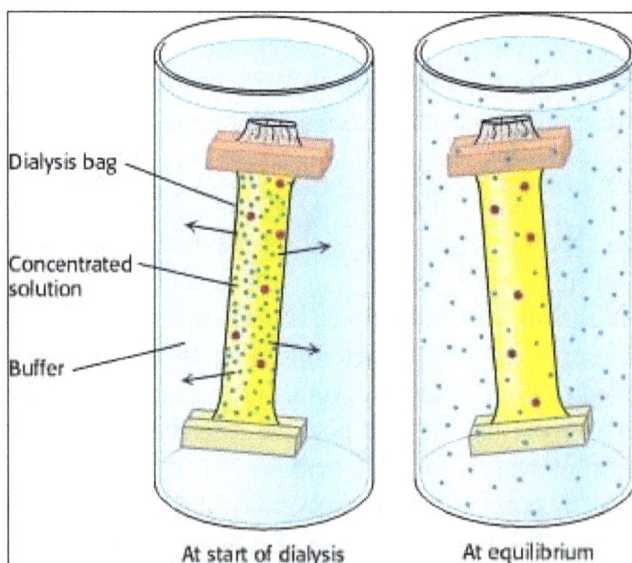

At start of dialysis                          At equilibrium

The scheme of dialysis. Enzyme molecules (red dots) are retained in the dialysis bag and separated from other smaller molecules (blue dots).

## 3. Polarity based separation

Like other proteins, enzymes can be separated on the basis of polarity, more specifically, their net charge, charge density, and hydrophobic interactions. In ion-exchange chromatography, a column of beads containing negatively or positively charged functional groups are used to separate enzymes. The cationic enzymes can be separated on anionic columns, and anionic enzymes on cationic column.

Electrophoresis is a procedure that uses an electrical field to cause permeation of ions through a solid or semi-solid matrix or surface resulting in separations on constituents on the basis of charge density. The most commonly used methods with a SDS-PAGE matrix are quite well standardized and do not differ much between labs. The distance a protein migrates in SDS-PAGE is inversely proportional to the log of its molecular radius, which is roughly proportional to molecular weight. Similarly, a matrix with gradient pH can be used in isoelectric focusing separation. A protein moves under the influence of an electrical field and stops upon reaching the pH which is the pI for the protein (net charge = 0). The matrix used can be liquid or a gel poured into either a cylindrical shape, or a flat plate.

Hydrophobic interaction chromatography (HIC) employs hydrophobic interactions to distinguish different enzymes, which are adsorbed on matrices such as octyl- or phenyl-Sepharose. A gradient of decreasing ionic strength, or possibly increasing non-polar solvent concentration can be used to elute the proteins, giving fractions that usually contain relatively high-pure enzymes. High-pressure liquid chromatography (HPLC) uses the same principle of separation of HIC, which is filled with more finely divided and tuned materials and thus allows more choices of eluents and results in better separation.

## 4. Affinity or ligand based purification

Affinity chromatography is another powerful and generally applicable means of purifying enzymes. This technique takes advantage of the high affinity of many enzymes for specific chemical groups. In general, affinity chromatography can be effectively used to isolate a protein that recognizes a certain group by (1) covalently attaching this group or a derivative of it to a column, (2) adding a mixture of proteins to this column, which is then washed with buffer to remove unbound proteins, and (3) eluting the desired protein by adding a high concentration of a soluble form of the affinity group or altering the conditions to decrease binding affinity. Affinity chromatography is most effective when the interaction of the enzyme and the molecule that is used as the bait is highly specific. A special example of ligand-affinity chromatography is the Ni-NTA (nickel – nitrolotriacetic acid-agaraose) affinity chromatography. This ligand binds tightly to a 6 amino acid peptide consisting only of histidines (His6). The cDNA sequence for His6 can be appended to the cDNA coding for a given recombinant protein, thus yielding a recombinant protein which contains a His-TAG. This allows the affinity-purification of such a protein using Ni-NTA without having to design a special ligand-affinity column.

Other forms of affinity chromatography include dye-ligand chromatography, immunoadsorption chromatography, and covalent chromatography.

After purification, the enzymes need to be concentrated, and sometimes lyophilized to give the pure, stable form distributed as the product or added into the final formulation. The following analysis and quality certification is necessary to confirm the enzyme is the desired one, with reasonable concentration, stability, and activity. The enrichment and certification requires experienced researchers to maintain the enzyme quality and choose the correct characterization method.

## Enzyme Purification by Electrophoresis

Electrophoresis is the motion of dispersed particles relative to a fluid under the influence of a spatially uniform electric field. It has become a basis of a number of analytical techniques, which can be used in enzyme separation by size, charge, or binding affinity. Several electrophoresis methods have been developed specifically.

1. Sodium dodecyl sulfate polyacrylamide gel electrophoresis (SDS-PAGE)

SDS-PAGE is a widely used method to analyze and isolate proteins. During the electrophoresis, proteins move towards to positive electrode, and the migration rate presents positive correlation to the molecular weight of the protein. SDS-PAGE is a convenient method for enzyme isolation and purity testing.

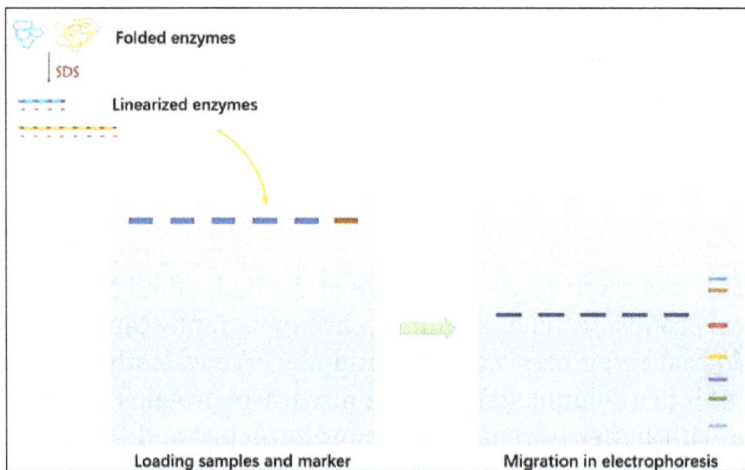

Enzymes isolation in SDS-PAGE system.

2. Isoelectric focusing (IEF)

Isoelectric focusing (IEF) is an electrophoretic method that allows the separation of amphoteric molecules (proteins, enzymes, peptides) in a pH gradient under the influence of an electric field according to their isoelectric points (pI). Different enzymes will stop at the positon where the pH equals to their pI. The purified enzymes are available after the separately elution.

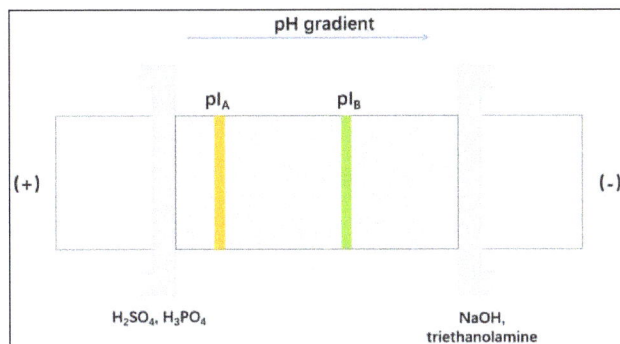

Enzymes isolation by IEF.

## 3. Affinity electrophoresis (AEP)

AEP technology is the combination of affinity chromatography and electrophoresis. During the electrophoresis, due to the affinity, the target molecule's migration rate decreases or remains at the origin. This method has become a simple and rapid tool for the analysis of binding affinities and inner reactions. It is commonly used in the separation, purification, and identification. The most two common approaches are lectin affinity electrophoresis (LAE) and capillary affinity electrophoresis (CAE).

# Single Substrate Reactions

Consider a simple single substrate reaction, where the free enzyme E binds to the substrate S to form a complex ES, the forms product P and then dissociates.

$$E + S \xleftrightarrow{K_1, K_{-1}} ES$$
$$ES \xleftrightarrow{K_2, K_{-2}} E + P$$

The dissociation of ES complex into free enzyme and product is the slowest and hence it is the rate-limiting step in the reaction. Throughout the reaction, the total concentration of the enzyme will be the sum of concentration of total free enzyme [E] and concentration of total enzyme bound with substrate [ES].

At pre-steady state, substrate concentration is more than that of the enzyme concentration. As reaction proceeds, the concentration of the enzyme substrate complex increases with time and reaches a steady state. After this stage, the concentration does not change further with time.

## Michaelis-Menten Kinetics and Briggs-Haldane Kinetics

The Michaelis-Menten model (1) is the one of the simplest and best-known approaches to enzyme kinetics. It takes the form of an equation relating reaction velocity to substrate concentration for a system where a substrate S binds reversibly to an enzyme E to form an enzyme-substrate complex ES, which then reacts irreversibly to generate a product P and to regenerate the free enzyme E. This system can be represented schematically as follows:

$$E + S \rightleftharpoons ES \rightarrow E + P$$

The Michaelis-Menten equation for this system is:

$$v = \frac{V_{max}[S]}{K_M + [S]}$$

Here, $V_{max}$ represents the maximum velocity achieved by the system, at maximum (saturating) substrate concentrations. $K_M$ (the Michaelis constant; sometimes represented as $K_S$ instead) is the substrate concentration at which the reaction velocity is 50% of the $V_{max}$. [S] is the concentration of the substrate S.

This is a plot of the Michaelis-Menten equation's predicted reaction velocity as a function of substrate concentration, with the significance of the kinetic parameters $V_{max}$ and $K_M$ graphically depicted.

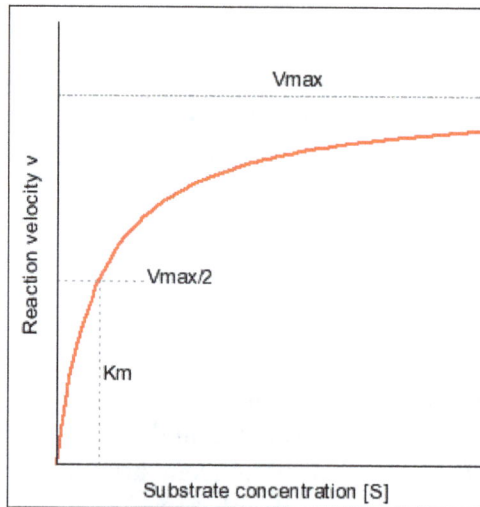

Michaelis-Menten model: enzyme kinetic reaction velocity

The best derivation of the Michaelis-Menten equation was provided by George Briggs and J.B.S. Haldane in 1925, and a version of it follows:

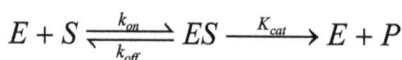

$$E + S \underset{k_{off}}{\overset{k_{on}}{\rightleftharpoons}} ES \xrightarrow{K_{cat}} E + P$$

For the scheme previously described, $k_{on}$ is the bimolecular association rate constant of enzyme-substrate binding; $k_{off}$ is the unimolecular rate constant of the ES complex dissociating to regenerate free enzyme and substrate; and $k_{cat}$ is the unimolecular rate constant of the ES complex dissociating to give free enzyme and product P. $k_{on}$ has units of concentration$^{-1}$time$^{-1}$, and $k_{off}$ and kcathave units of time$^{-1}$. Also, by definition the dissociation binding constant of the ES complex, $K_D$ is given by $k_{off}/k_{on}$ (and so has units of concentration). Once these rate constants have been defined, we can write equations for the rates of change of all the chemical species in the system:

$$\frac{d[S]}{dt} = -k_{on}[E][S] + k_{off}[ES]$$

$$\frac{d[E]}{dt} = -k_{on}[E][S] + (k_{off} + k_{cat})[ES]$$

$$\frac{d[P]}{dt} = k_{cat}[ES]$$

$$\frac{d[ES]}{dt} = k_{on}[E][S] - (k_{off} + k_{cat})[ES]$$

The last of these equations – describing the rate of change of the ES complex – is the most important for our purposes. In most systems, the ES concentration will rapidly approach a steady-state – that is, after an initial burst phase, its concentration will not change appreciably until a significant amount of substrate has been consumed. This steady-state approximation is the first important assumption involved in Briggs and Haldane's derivation. This is also the reason that well-designed experiments measure reaction velocity only in regimes where product formation is linear with time. To study initial reaction velocities, we can assume that [ES] is constant:

$$\frac{d[ES]}{dt} = 0$$

$$\Rightarrow k_{on}[E][S] = (k_{off} + k_{cat})[E][S]$$

In order to determine the rate of product formation ($d[P]/dt = k_{cat}[ES]$), we need to rearrange the equation above to calculate [ES]. We know that the free enzyme concentration [E] is equal to the total enzyme concentration [$E_T$] minus [ES]. Making these substitutions gives us:

$$k_{on}([E_T] - [ES])[S] = k_{off} + k_{cat}[ES]$$
$$k_{on}[E_T][S] - k_{on}[ES][S] = k_{off} + k_{cat}[ES]$$
$$k_{on}[E_T][S] = k_{off} + k_{cat}[ES] + k_{on}[ES][S]$$

We now make a couple of substitutions to arrive at the familiar form of the

Michaelis-Menten equation. Since $V_{max}$ is the reaction velocity at saturating substrate concentration, it is equal to $k_{cat}$ [ES] when [ES] = [E$_T$]. We also define $K_M$ in terms of the rate constants as follows:

$$V_{max} = k_{cat}[E_T]; \quad K_M = \frac{k_{off} + k_{cat}}{k_{on}}$$

$$v = \frac{V_{max}[S]}{K_M + [S]}$$

The S represents the free substrate concentration, but typically is assumed to be close to the total substrate concentration present in the system. This second assumption is the free ligand approximation, and is valid as long the total enzyme concentration is well below the KM of the system. If this condition is not met (for instance, with a very high-affinity substrate), then the quadratic (or 'Morrison') equation must be used instead.

Comparing $K_M$ [= ($k_{off}$ + $k_{cat}$)/$k_{on}$] and $K_D$ [= $k_{off}$/$k_{on}$], it is obvious that $K_M$ must always be greater than $K_D$. Michaelis and Menten assumed that substrate binding and dissociation occurred much more rapidly than product formation ($k_{cat}$ << $k_{off}$, the rapid equilibrium approximation), and that therefore the $K_M$ would be very close to the $K_D$. The larger the $k_{cat}$ is relative to $k_{off}$, the greater the difference between $K_D$ and $K_M$. Briggs and Haldane made no assumptions about the relative values of $k_{off}$ and $k_{cat}$, and so Michaelis-Menten kinetics are a special case of Briggs-Haldane kinetics. The opposite extreme, where $k_{cat}$ >> $k_{off}$, is called Van Slyke-Cullen behavior.

## Quadratic Velocity Equation for Tight-Binding Substrates

Three assumptions are implicit in Michaelis-Menten kinetics: the steady-state approximation, the free ligand approximation and the rapid equilibrium approximation. (The Briggs-Haldane approach frees us from the last of these three.)

$$v = \frac{V_{max}[S]}{K_M + [S]}$$

Now consider the second assumption in the list: the free ligand approximation. this approximation states that the free substrate concentration (the [S] in the Michaelis-Menten equation) is close to the total substrate concentration in the system – which is, in fact, the true independent variable in most experimental set-ups. we know how much substrate we've added to reaction mixture, but we don't know a priori how much of that substrate remains free in solution at steady-state. The free ligand approximation does not hold when a substrate's $K_M$ is lower than the total enzyme concentration, because at low substrate concentrations, a significant fraction of substrate will be bound to the enzyme.

$$E + S \underset{\text{off}}{\overset{\text{on}}{\rightleftharpoons}} ES \overset{\text{cat}}{\longrightarrow} E + P$$

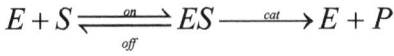

For such unusual cases, we derive a kinetic equation from the scheme above without resorting to the free ligand approximation. The rates of change of the various species are given by four differential equations:

$$\frac{d[S]}{dt} = -k_{on}[E][S] + k_{off}[ES]$$

$$\frac{d[E]}{dt} = -k_{on}[E][S] + (k_{off} + k_{cat})[ES]$$

$$\frac{d[P]}{dt} = k_{cat}[ES]$$

$$\frac{d[ES]}{dt} = k_{on}[E][S] - (k_{off} + k_{cat})[ES]$$

We can assume that ES levels achieve steady-state:

$$\frac{d[ES]}{dt} = \quad \Rightarrow k_{on}[E][S] = (k_{off} + k_{cat})[ES]$$

Given that free enzyme concentration [E] equals total enzyme concentration $[E_T]$ minus [ES] and that [S] equals total enzyme concentration, we have:

$$k_{on}([E_T] - [ES])([S_T] - [ES]) = (k_{off} + k_{cat})[ES]$$

A little algebra gives us:

$$([E_T] - [ES])([S_T] - [ES]) = \frac{k_{off} + k_{cat}}{k_{on}}[ES] = K_M[ES]$$

$$[E_T][ES] - ([E_T] + [S_T])[ES] + [ES]^2 = K_M[ES]$$

$$[ES]^2 - ([E_T] + [S_T] + [K_M])[ES] + [E_T][S_T] = 0$$

This is a quadratic equation of the form $a^2x + bx + c = 0$, whose roots are given by:

$$\frac{\left(-b \pm \sqrt{b^2 - 4ac}\right)}{2a}$$

Making the following substitutions and choosing the relevant (saturable) root gives us:

$$a = 1$$
$$b = -([E_T] + [S_T] + K_M)$$
$$c = [E_T][S_T]$$

$$[ES] = \frac{([E_T]+[S_T]+K_M) - \sqrt{([E_T]+[S_T]+K_M)^2 - 4[E_T][S_T]}}{2}$$

$$v = k_{cat}[ES] = \frac{V_{max}}{[E_T]}[ES]$$

$$v = V_{max}\frac{([E_T]+[S_T]+K_M) - \sqrt{([E_T]+[S_T]+K_M)^2 - 4[E_T][S_T]}}{2[E_T]}$$

This is the quadratic velocity equation, sometimes also called the tight-binding equation or the Morrison equation (4). As the $K_M$ becomes larger than $[E_T]$, the curve described by the quadratic equation approaches the hyperbola described by the Michaelis-Menten equation. As the $K_M$ becomes lower, the inflection point of the curve (located where $[S_T] = [E_T]$) becomes progressively sharper. This is illustrated in the plot below, where the $K_M$ varies from 5 times the total enzyme concentration ($[E_T]$ = 1 µM) to one-hundredth of $[E_T]$. Differences in affinity between very tight-binding substrates are reflected in the sharpness of the inflection, meaning that $K_M$ values recovered from fits to experimental data are disproportionately sensitive to error in the data points surrounding the inflection point. As a rule of thumb, the quadratic equation should be used in preference to the Michaelis-Menten equation whenever the $K_M$ is less than five-fold larger than $[E_T]$.

Derivations of this type that do not use the free ligand approximation grow rapidly more complicated for more complex kinetic systems. For anything more complicated than a one-site model, it is better to use kinetic simulations as detailed in part II of this guide.

## Multiple Binding: Sequential Models

Early work in this regard was carried out by Adair and Pauling, operating under the rapid equilibrium approximation. this assumption states that all substrate binding and dissociation steps happen much more rapidly than catalytically productive steps., King and Altman showed how to solve any kinetic system without resorting to this approximation, but this is unnecessarily complicated for our purposes. Even if the rapid equilibrium approximation does not hold for our system, the forms of the equations we derive will not change – instead, the significance of the various $K_M$ values will be different. This is analogous to the difference between $K_M$ (or $K_D$) in the Michaelis-Menten model and $K_M$ in the Briggs-Haldane model.

We begin with the simplest model of multiple binding: a two-site sequential model. Here, an enzyme E can bind a single molecule of substrate S to form a singly-occupied complex ES with equilibrium dissociation constant $K_{D1}$. ES can either react irreversibly to form product P with rate $k_{cat1}$, or can bind a second substrate molecule S to form a doubly-occupied complex ESS with dissociation constant $K_{D2}$. In turn, ESS can irreversibly form P with rate $k_{cat2}$. This is represented schematically.

$$E \underset{K_{D1}}{\overset{S}{\rightleftharpoons}} ES \xrightarrow{k_{cat}} E + P$$

$$ES \overset{S}{\underset{K_{D2}}{\Updownarrow}}$$

$$ESS \xrightarrow{k_{cat}} ES + P$$

In order to calculate the reaction velocity for this system, we need to know the relative concentrations of the active species ES and ESS. We will describe a straightforward derivation of these quantities, and then a shortcut that allows one to quickly solve more complicated kinetic systems.

From the definitions of the two dissociation constants, we have:

$$K_{D_1} = \frac{[E][S]}{[ES]} \Rightarrow [ES] = \frac{[E][S]}{K_{D_1}}$$

$$K_{D_2} = \frac{[E][S]}{[ESS]} \Rightarrow [ESS] = \frac{[ES][S]}{K_{D_2}} = \frac{[E][S]^2}{K_{D_1} K_{D_2}}$$

Relative to the total enzyme concentration, $E_T$:

$$\frac{[ES]}{[E_T]} = \frac{[ES]}{[E]+[ES]+[ESS]} = \frac{\dfrac{[E][S]}{K_{D_1}}}{[E] + \dfrac{[E][S]}{K_{D_1}} + \dfrac{[E][S]^2}{K_{D_1} K_{D_2}}} = \frac{\dfrac{[S]}{K_{D_1}}}{1 + \dfrac{[S]}{K_{D_1}} + \dfrac{[S]^2}{K_{D_1} K_{D_2}}}$$

$$\frac{[ESS]}{[E_T]} = \frac{[ESS]}{[E]+[ES]+[ESS]} = \frac{\dfrac{[E][S]^2}{K_{D_1} K_{D_2}}}{[E] + \dfrac{[E][S]}{K_{D_1}} + \dfrac{[E][S]^2}{K_{D_1} K_{D_2}}} = \frac{\dfrac{[S]^2}{K_{D_1} K_{D_2}}}{1 + \dfrac{[S]}{K_{D_1}} + \dfrac{[S]^2}{K_{D_1} K_{D_2}}}$$

The total reaction velocity for this system is given by (where $V_{max1} = k_{cat1}[E_T]$ and $V_{max2} = k_{cat2}[E_T]$):

$$v = \frac{V_{max_1} \dfrac{[S]}{K_{D_1}} + V_{max_2} \dfrac{[S]^2}{K_{D_1} K_{D_2}}}{1 + \dfrac{[S]}{K_{D_1}} + \dfrac{[S]^2}{K_{D_1} K_{D_2}}}$$

This graph shows the relative concentrations of the two active species as a function of substrate concentration, when both KDs are set to 10 µM. The initial increase and subsequent decrease in the level of ES, while the concentration curve for ESS is sigmoidal. Therefore, if $V_{max1}$ (for ES) is much greater than $V_{max2}$ (for ESS) then the system will exhibit substrate inhibition kinetics; if $V_{max2}$ is much greater than $V_{max1}$ then the system will exhibit sigmoidal kinetics.

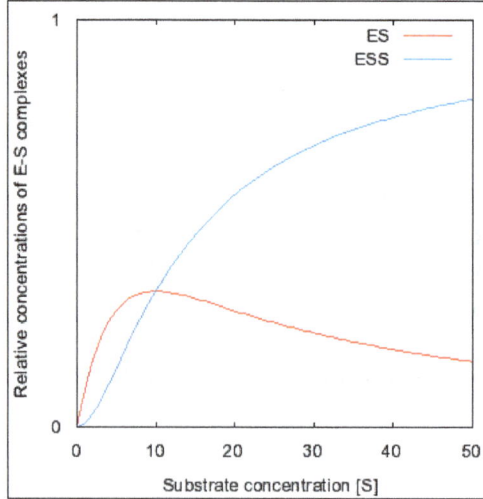

Figure $K_D s = 10\mu M$

There is a quicker way to calculate reaction velocity equations of this type. To return to the scheme of a two-site sequential model:

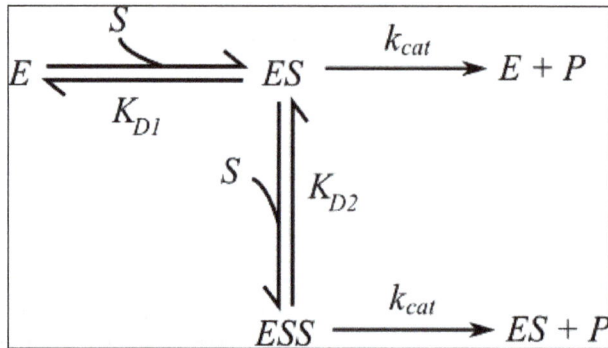

Each species in the system has a relative concentration term that appears in the denominator of the velocity equation, while terms from all active species appear in the numerator. To find the relative concentration term of any species, simply work your way backwards from that species to the free enzyme; the specific term is equal to the product of all the substrate molecules that bind, divided by all the equilibrium constants in the chain. So the specific term for ES is $[S]/K_{D1}$ and the term for ESS is $[S]^2/K_{D1}K_{D2}$. The specific term for free enzyme is always 1.

The denominator of the velocity equation is the sum of all specific terms: $1 + [S]/K_{D1} + [S]^2/K_{D1}K_{D2}$; the numerator is the sum of specific terms for catalytically active species, each weighted by their respective $V_{max}$ values – and voila.

$$v = \cfrac{V_{max_1}\cfrac{[S]}{K_{D_1}} + V_{max_2}\cfrac{[S]^2}{K_{D_1}K_{D_2}}}{1 + \cfrac{[S]}{K_{D_1}} + \cfrac{[S]^2}{K_{D_1}K_{D_2}}}$$

## Multiple Binding: Random-ordered Models

In a two-distinct-site random-ordered model, there are two substrate binding sites on the enzyme, each with their own $K_D$s. Once either site is occupied, the substrate can bind to the other site with an altered $K_D$. If the substrate binding at one site enhances the affinity of binding at another site, then the enzyme exhibits positive cooperativity. If the reverse is true (binding at one site decreases the affinity of the other site), the enzyme shows negative cooperativity.

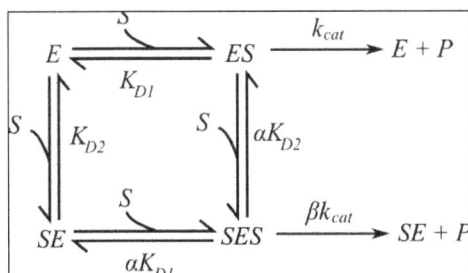

In the scheme above, ES and SE are the two singly-occupied complexes and SES is the doubly-occupied complex. ES and SES are both productive: the parameter $\beta$ reflects how productive the SES complex is relative to ES, in that values below 1 imply substrate inhibition and values above 1 imply activation. The cooperativity parameter $\alpha$ shows how binding at one site affects the affinity at the other: values below 1 indicate increased affinity (positive cooperativity) and values above 1 indicate negative cooperativity.

Extending the specific terms for ES, SE and SES are $[S]/K_{D1}$, $[S]/K_{D2}$, and $[S]^2/\alpha K_{D1}K_{D2}$ respectively. Therefore, the velocity equation for the system is:

$$v = \cfrac{V_{max}\cfrac{[S]}{K_{D1}} + \beta V_{max}\cfrac{[S]^2}{\alpha K_{D1}K_{D2}}}{1 + \cfrac{[S]}{K_{D1}} + \cfrac{[S]}{K_{D2}} + \cfrac{[S]^2}{\alpha K_{D1}K_{D2}}}$$

If one binding site is much higher-affinity than the other, the specific term associated with the weaker-binding (higher $K_D$) site is negligible and can therefore be omitted from this equation, which then resembles the velocity equation previously derived for the two-site sequential model. In other words, sequential binding models are a special case of random-ordered models where one binding site is much higher-affinity than the other ($K_{D_1} >> K_{D_2}$ or $K_{D_1} << K_{D_2}$).

Although sequential models are less general than random-ordered models, they have fewer free parameters and so are easier to use when fitting experimental data. This means that, in cases where they are applicable, sequential models are often a better choice than random-ordered models when it comes to analyzing experimental data.

## Homotropic Allosterism

Here, we use the random-ordered binding model to describe how common forms of atypical (non-hyperbolic) allosteric behavior may arise. We focus here on homotropic allosterism, where multiple molecules of a singly chemical species can bind the enzyme.

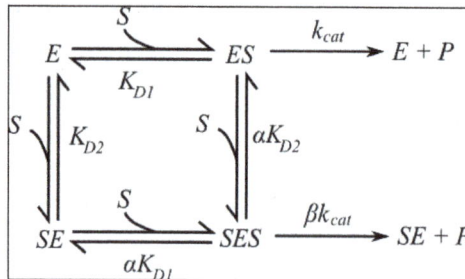

We know that the scheme above, showing a two-site random-ordered model, is described by the following velocity equation:

$$v = \frac{V_{max}\dfrac{[S]}{K_{D1}} + \beta V_{max}\dfrac{[S]^2}{\alpha K_{D1} K_{D2}}}{1 + \dfrac{[S]}{K_{D1}} + \dfrac{[S]}{K_{D2}} + \dfrac{[S]^2}{\alpha K_{D1} K_{D2}}}$$

Different combinations of the parameters $\alpha$ and $\beta$ can produce hyperbolic or allosteric behavior. For instance, if we set $K_{D1} = K_{D2} = 5\ \mu M$ with $\alpha = 1$ (no cooperativity in binding) and $\beta = 1$ (ES and SES have identical activity), the overall velocity equation corresponds to a hyperbola with an apparent $K_M$ of 5 $\mu M$.

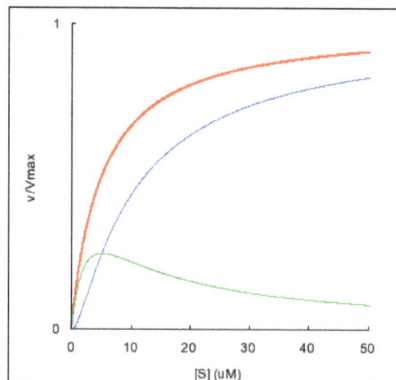

Figure $K_{D1} = K_{D2} = 5\ \mu M$, $\alpha = 1$, $\beta = 1$

Total velocity is shown in red, with contributions from ES in green and SES in blue. In the absence of binding cooperativity and substrate activation, with equal $K_D$s for both binding sites, the velocity curve appears hyperbolic.

Sigmoidal kinetics can be caused by positive binding cooperativity ($\alpha < 1$) or by substrate activation ($\beta > 1$).

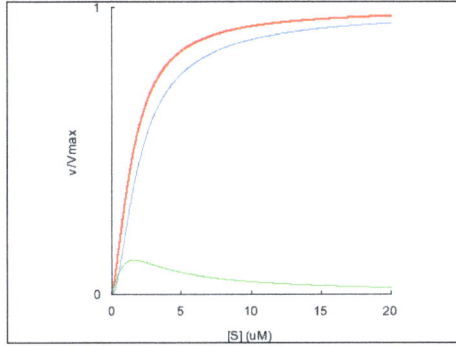

$K_{D1} = K_{D2} = 5\ \mu M,\ \alpha = 0.1,\ \beta = 1$

Sigmoidal kinetics can result from positive binding cooperativity.

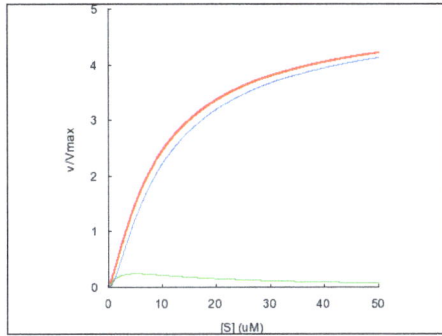

$K_{D1} = K_{D2} = 5\ \mu M,\ \alpha = 1,\ \beta = 5$

Sigmoidal kinetics can also result from substrate activation, even without binding cooperativity.

- Biphasic kinetics can result from negative binding cooperativity ($\alpha < 1$).

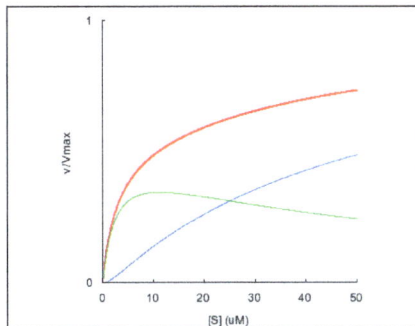

$K_{D1} = K_{D2} = 5\ \mu M,\ \alpha = 10,\ \beta = 1$

- Substrate inhibition occurs when SES is less active than ES ($\beta < 1$).

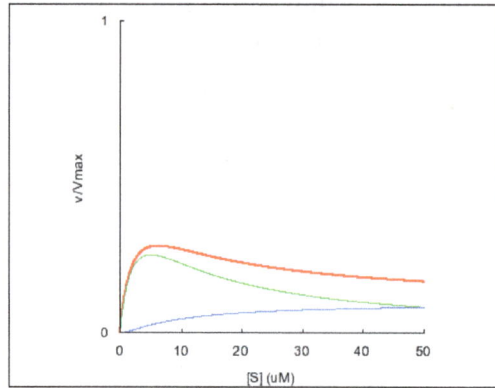

$$K_{D_1} = K_{D_2} = 5 \, \mu M, \, \alpha = 1, \, \beta = 0.1$$

## Inhibition and Allosterism

Random-ordered models can easily be adapted to describe many common modes of enzyme inhibition and activation by chemical species different from the substrate. The following scheme is a generalized model of inhibition that can describe competitive, uncompetitive, mixed and non-competitive inhibition, as well as heterotropic activation.

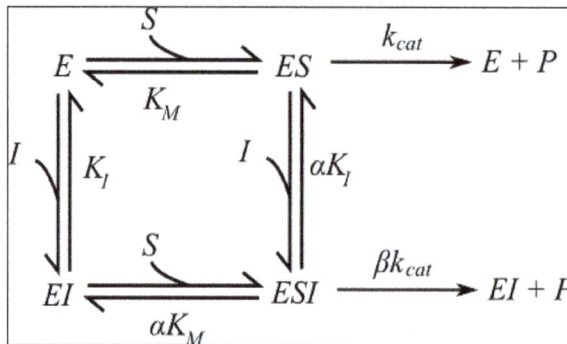

The specific terms for ES, EI and ESI are $[S]/K_M$, $[I]/K_I$, and $[S][I]/\alpha K_M K_I$ respectively. This leads to the following velocity equation for the general case:

$$v = \frac{V_{max} \dfrac{[S]}{K_M} + \beta V_{max} \dfrac{[S][I]}{\alpha K_M K_I}}{1 + \dfrac{[S]}{K_M} + \dfrac{[I]}{K_I} + \dfrac{[S][I]}{\alpha K_M K_I}}$$

This equation represents a 3-D surface with [S] and [I] as independent variables.

The parameter β describes the extent of inhibition (when $\beta < 1$) or the extent of activation (when $\beta > 1$). This system approaches competitive inhibition (where inhibitor binds to the free enzyme but not the ES complex) when $\alpha \gg 1$ and so $K_I \ll \alpha K_I$. Uncompetitive inhibition, where I binds ES but not E, is achieved when $K_I \gg \alpha K_I$. Non-competitive inhibition occurs when $\alpha = 1$, and I binds E and ES with equal affinities. All other combinations of parameters represent mixed inhibition.

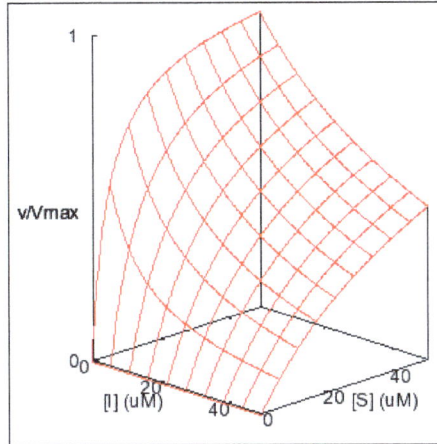

Competitive Inhibition: $K_M = 5\ \mu M$, $K_I = 5\ \mu M$, $\alpha = 1000$, $\beta = 0$

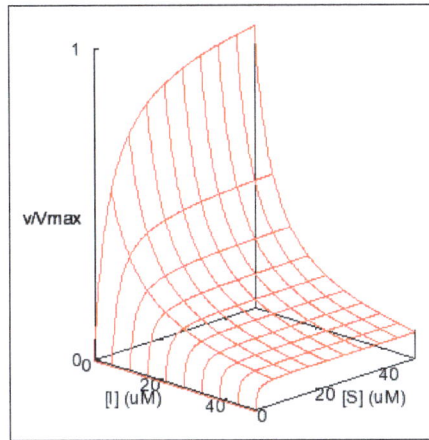

Uncompetitive Inhibition: $K_M = 5\ \mu M$, $K_I = 5000\ \mu M$, $\alpha = 0.001$, $\beta = 0$

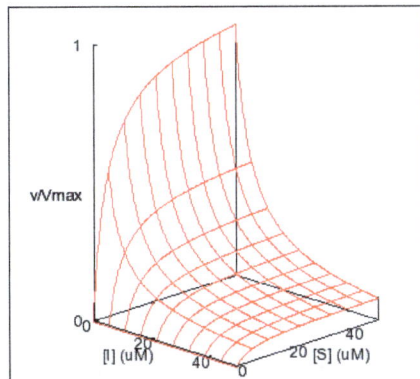

Non-competitive Inhibition: $K_M = 5\ \mu M$, $K_I = 5\ \mu M$, $\alpha = 1$, $\beta = 0$

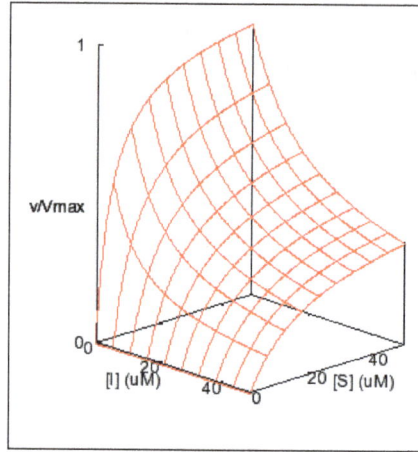

Partial Mixed Inhibition $K_M = 5$ μM, $K_I = 5$ μM, $\alpha = 5$, $\beta = 0.2$

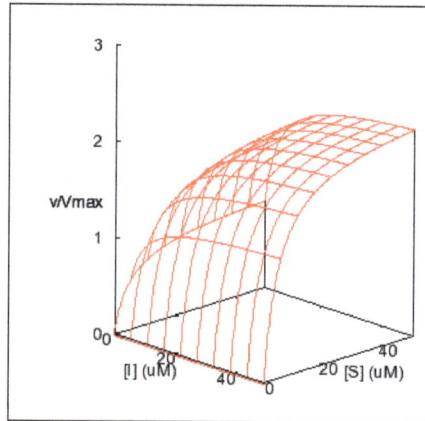

Activation $K_M = 5$ μM, $K_I = 5$ μM, $\alpha = 0.2$, $\beta = 2.5$

More complex models that incorporate multiple binding sites for substrates and inhibitors can be constructed using the techniques outlined here, but this is left as an exercise for the reader.

Translating Between Velocity Equations and Binding Equations many experiments focus not on catalysis but on equilibrium binding, using absorbance or fluorescence changes induced by substrate binding as observable signals. The strategies we used to derive velocity equations for kinetic models are easily adapted to derive binding equations for equilibrium models. Basically, all $K_M$s are replaced by $K_D$s and all $V_{max}$ values are replaced by appropriate scaling factors for the observed signals.

For example, consider a two-site sequential equilibrium binding model:

$$E \underset{K_{D1}}{\overset{+S}{\rightleftharpoons}} ES \underset{K_{D2}}{\overset{+S}{\rightleftharpoons}} ESS$$

If the ES and ESS species have differential responses to the experimental technique used, they will each have a maximal response $B_{max1}$ and $B_{max2}$. Then the equilibrium binding equation (where B is the observed signal) is:

$$B = \frac{B_{max1}\dfrac{[S]}{k_{D1}} + B_{max2}\dfrac{[S]^2}{K_{D1}K_{D2}}}{1 + \dfrac{[S]}{k_{D1}} + \dfrac{[S]^2}{K_{D1}K_{D2}}}$$

# Enzyme Substrate Complex

The enzyme substrate complex is a temporary molecule formed when an enzyme comes into perfect contact with its substrate. Without its substrate an enzyme is a slightly different shape. The substrate causes a conformational change, or shape change, when the substrate enters the active site. The active site is the area of the enzyme capable of forming weak bonds with the substrate. This shape change can force two or more substrate molecules together, or split individual molecules into smaller parts. Most reactions that cells use to stay alive require the actions of enzymes to happen fast enough to be useful. These enzymes are directly coded for in the DNA of the organism.

The enzyme substrate complex is extremely important for a number of reasons. First, the enzyme substrate complex is only temporary. This means that once the substrate has changed, it can no longer bind to the enzyme. The products are released and the enzyme is ready for another substrate molecule. A single enzyme can operate repeatedly millions of times, meaning only a small amount of enzyme is needed in each cell.

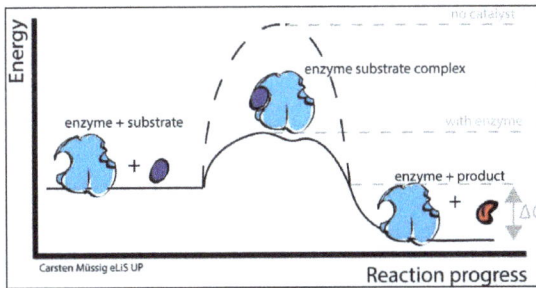

Enzyme action

Enzymes are complex molecules, like little machines meant for one purpose. Built out of a chain of amino acids, this long string experiences interactions between the different amino acids and twists and turns into complex structures. These structures can operate like hinges, wedges, and all sorts of other shapes intended to speed certain reactions. Different mutations give rise to slightly different forms of enzyme. In mutations that are beneficial to the organism, the enzyme substrate complex is changed in a way that effects the output of product or the function of the enzyme as a whole. This change in the organism is only beneficial if it somehow helps the organism reproduce more.

Enzymes are usually named after the substrate that they work on, and have the -ase suffix to designate they are enzymes. Each enzyme has a certain specificity for the substrate it works on, which determines which molecules they can bind to. Some molecules that are similar in structure to the substrate may get stuck in the active site, because they cannot undergo the reaction intended by the enzyme. In this warped enzyme substrate complex, the competitive inhibitor binds to the enzyme and inhibits its further action. Other inhibitors do not copy the substrate, but modify the enzyme in other ways so the enzyme substrate complex cannot be formed.

## Examples of Enzyme Substrate Complex

### Amylase and Amylose

Amylose is a complex sugar produced by plants. In our saliva is an enzyme, amylase, used to break amylose apart. Amylase uses one substrate molecule of amylose and a cofactor of one water molecule to produce an enzyme substrate complex. The complex severely reduces the amount of energy required to start the reaction, which increases the time in which it happens. A typical sugar molecule would take millions of years to break apart, were it not for the actions of enzymes such as amylase.

In fact, enzymes are so important in digesting the foods we eat that our body produces an enzyme for almost every type of food the body is evolutionarily prepared to consume. New foods are poorly processed, because the enzymes have not had time to adjust their efficiency. For instance, the modern diet of processed foods is leading to an obesity epidemic because the process foods are rich in easily accessible nutrients, but only to the pathways that are used to storing fat. As a result, much of the population experiences weight related illnesses. Many nutritionists are pushing for more natural, whole-food, plant based diets that tend to support the enzymes our bodies have naturally developed.

### Allosteric Regulation in Enzymes

Enzymes have an active site that provides a unique chemical environment, made up of certain amino acid R groups (residues) in a particular orientations and distance from one another. This unique environment is well-suited to convert particular chemical reactants for that enzyme, called substrates, into unstable intermediates (transition states). Enzymes and substrates are thought to bind with an "induced fit", which means that enzymes and substrates undergo slight conformational adjustments upon substrate contact, leading to binding. This subtle change in enzyme shape allows the enzyme to rapidly bind potential substrates in an "open" conformation" and then generate a tighter "closed" catalytically active alternative conformation only when the correct substrate is correctly aligned in the active site.

Enzymes bind to substrates and can potentially catalyze reactions in four different ways (which maybe act together in a single enzyme): bringing substrates together in an optimal orientation, compromising the bond structures of substrates so that bonds can be more easily broken, providing optimal environmental conditions (often local pH) for a reaction to occur, and participating directly in their chemical reaction by forming transient covalent bonds with their substrates.

Enzyme action must be regulated so that in a given cell at a given time, the desired reactions are being catalyzed and the undesired reactions are not. Inhibition and activation of enzymes via other molecules are important ways that enzymes are regulated. Inhibitors can act competitively, noncompetitively, or allosterically (allo (other) steric (form)); noncompetitive inhibitors are usually allosteric. Activators can also enhance the function of enzymes allosterically. The most common method by which cells regulate the enzymes in metabolic pathways is through feedback inhibition. During feedback inhibition, the products of a metabolic pathway serve as inhibitors (usually allosteric) of one or more of the enzymes (usually the first committed enzyme of the pathway) involved in the pathway that produces them.

## Enzyme Active Site and Substrate Specificity

The chemical reactants to which an enzyme binds are the enzyme's substrates. There may be one or more substrates, depending on the particular chemical reaction. In some reactions, a single-reactant substrate is broken down into multiple products. In others, two substrates may come together to create one larger molecule. Two reactants might also enter a reaction, both become modified, and leave the reaction as two products. The location within the enzyme where the substrate binds is called the enzyme's active site. Since enzymes are proteins, there is a unique combination of amino acid R groups within the active site. Each amino acid side-chain is characterized by different properties. The unique combination of amino acids, their positions, sequences, structures, and properties, creates a very specific chemical environment within the active site. This specific environment is suited to bind, albeit briefly, to a specific chemical substrate (or substrates). Due to this jigsaw puzzle-like match between an enzyme and its substrates, enzymes can be extremely specific in their choice of substrates. The "best fit" between an enzyme and its substrates results from the their respective shapes and the chemical complementarity of the functional groups on each binding partner.

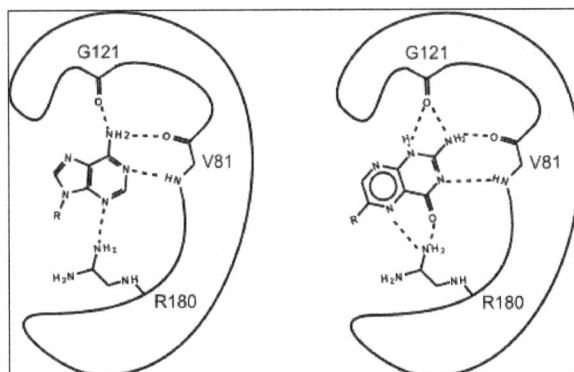

This is an enzyme with two different substrates bound in the active site (here conveniently squished down to 2 dimensions). The enzymes are represented as blobs, except for the active site which identifies three amino acids located in the active site (and shows the R group for one of them). The R group of R180 is interacting with the substrates through hydrogen bonding (represented as dashed lines), as are some groups in the peptide backbone. Amino acid positions are denoted a single letter code for the amino acid followed immediately by "position of the amino acid vs. the N terminal end". For example "R180" means an R (arginine) is the 180th amino acid from the N terminus.

At this point in the class you should be familiar with the chemical characteristics (charge, polarity, hydrophobicity) of the functional groups. For example, the R group of R180 in the enzyme depicted above is the amino acid Arginine (arginine's single letter code happens to be R, which is a little confusing in this context) and R180's R group consists of several "amino" functional groups. An amino functional group contains a nitrogen (N) and hydrogen (H) atoms. Nitrogen is more electronegative than hydrogen so the covalent bond between N-H is a polar covalent bond. The hydrogen atoms in this bond will have a partial positive charge, and the nitrogen atom will have a partial negative charge. This allows amino groups to form hydrogen bonds with other polar compounds. Likewise, the backbone carbonyl oxygens of Valine (V81) and Glycine (G121) the backbone amino hydrogen of V81 are depicted engaged in hydrogen bonds with the small molecule substrate.

## Structural Instability of Enzymes

The fact that active sites are so well-suited to provide specific environmental conditions also means that they are subject to influences by the local environment. It is true that increasing the environmental temperature generally increases reaction rates, enzyme-catalyzed or otherwise. However, increasing or decreasing the temperature outside of an optimal range can affect chemical bonds within the active site in such a way that they are less well suited to bind substrates. High temperatures will eventually cause enzymes, like other biological molecules, to denature, a process that changes the natural properties of a substance. Likewise, the pH of the local environment can also affect enzyme function. Active site amino acid residues have their own acidic or basic properties that are optimal for catalysis. These residues are sensitive to changes in pH that can impair the way substrate molecules bind, because the charges on the R groups, and therefore both ionic and H-bonding interactions can change with pH. Enzymes are suited to function best within a certain pH range, and, as with temperature, extreme pH values (acidic or basic) of the environment can cause enzymes to denature.

The process where enzymes denature usually starts with the unwinding of the tertiary structure through destabilization of the bonds holding the tertiary structure together. Hydrogen bonds, ionic bonds and covalent bonds (disulfide bridges and peptide bonds) can all be disrupted by large changes in temperate and pH. Using the chart of enzyme activity and temperature below, make an energy story for the red enzyme. Explain what might be happening from temperature 37 °C to 95 °C.

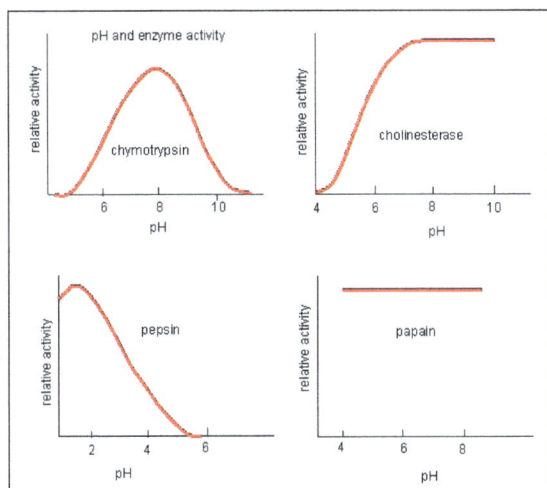

Enzymes have an optimal pH.

The pH at which the enzyme is most active will be the pH where the active site R groups are protonated/deprotonated such that the substrate can enter the active site and the initial step in the reaction can begin. Some enzymes require a very low pH (acidic) to be completely active. In the human body, these enzymes are most likely located in the stomach, or located in lysosomes (a cellular organelle used to digest large compounds inside the cell).

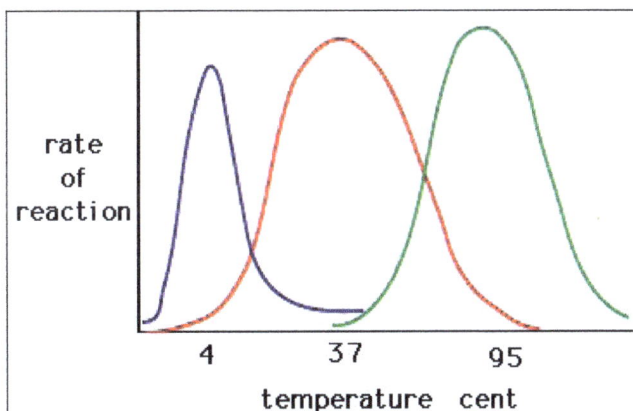

Enzymes have an optimal temperature.

The temperature at which the enzyme is most active will usually be the temperature where the structure of the enzyme is stable or uncompromised. Some enzymes require a specific temperature to remain active and not denature.

## Induced Fit and Enzyme Function

For many years, scientists thought that enzyme-substrate binding took place in a simple "lock-and-key" fashion. This model asserted that the enzyme and substrate fit together perfectly in one instantaneous step. However, current research supports a more refined view called induced fit. The induced-fit model expands upon the lock-and-key model by describing a more dynamic interaction between enzyme and substrate. As the enzyme and substrate come together, their interaction causes a mild shift in the enzyme's structure that confirms an more productive binding arrangement between the enzyme and the transition state of the substrate. This energetically favorable binding maximizes the enzyme's ability to catalyze its reaction.

When an enzyme binds its substrate, an enzyme-substrate complex is formed. This complex lowers the activation energy of the reaction and promotes its rapid progression in one of many ways. On a basic level, enzymes promote chemical reactions that involve more than one substrate by bringing the substrates together in an optimal orientation. The appropriate region (atoms and bonds) of one molecule is juxtaposed to the appropriate region of the other molecule with which it must react. Another way in which enzymes promote the reaction of their substrates is by creating an energetically favorable environment within the active site for the reaction to occur. Certain chemical reactions might proceed best in a slightly acidic or non-polar environment. The chemical properties that emerge from the particular arrangement of amino acid residues within an active site create the energetically favorable environment for an enzyme's specific substrates to react.

The activation energy required for many reactions includes the energy involved in slightly contorting chemical bonds so that they can more easily react. Enzymatic action can aid this process. The enzyme-substrate complex can lower the activation energy by contorting substrate molecules in such a way as to facilitate bond-breaking. Finally, enzymes can also lower activation energies by taking part in the chemical reaction itself. The amino acid residues can provide certain ions or chemical groups that actually form covalent bonds with substrate molecules as a necessary step of the reaction process. In these cases, it is important to remember that the enzyme will always return to its original state at the completion of the reaction. One of the hallmark properties of enzymes is that they remain ultimately unchanged by the reactions they catalyze. After an enzyme is done catalyzing a reaction, it releases its product(s).

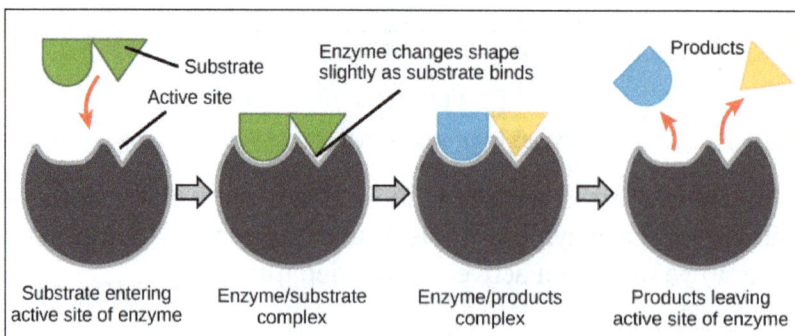

According to the induced-fit model, both enzyme and substrate undergo dynamic conformational changes upon binding. The enzyme contorts the substrate into its transition state, thereby increasing the rate of the reaction.

## Enzyme Regulation

Cellular needs and conditions vary from cell to cell, and change within individual cells over time. The required enzymes and energetic demands of stomach cells are different from those of fat storage cells, skin cells, blood cells, and nerve cells. Furthermore, a digestive cell works much harder to process and break down nutrients during the time that closely follows a meal compared with many hours after a meal. As these cellular demands and conditions vary, so do the needed amounts and functionality of different enzymes.

## Regulation of Enzymes by Molecules

Enzymes can be regulated in ways that either promote or reduce their activity. There are many different kinds of molecules that inhibit or promote enzyme function, and various mechanisms exist for doing so. In some cases of enzyme inhibition, for example, an inhibitor molecule is similar enough to a substrate that it can bind to the active site and simply block the substrate from binding. When this happens, the enzyme is inhibited through competitive inhibition, because an inhibitor molecule competes with the substrate for active site binding. On the other hand, in noncompetitive inhibition, an inhibitor molecule binds to the enzyme in a location other than an allosteric site and still manages to block substrate binding to the active site.

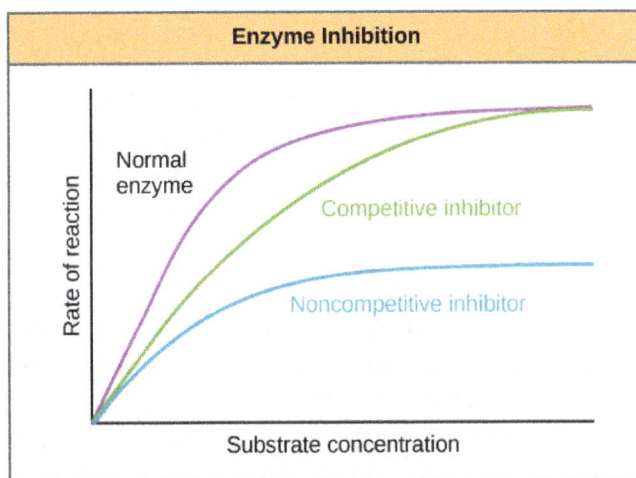

Competitive and noncompetitive inhibition affect the rate of reaction differently.
Competitive inhibitors affect the initial rate but do not affect the maximal rate,
whereas noncompetitive inhibitors affect the maximal rate.

Some inhibitor molecules bind to enzymes in a location where their binding induces a conformational change that reduces the affinity of the enzyme for its substrate. This type of inhibition is called allosteric inhibition. Most allosterically regulated enzymes are made up of more than one polypeptide, meaning that they have more than one protein subunit. When an allosteric inhibitor binds to an enzyme, all active sites on the protein subunits are changed slightly such that they bind their substrates with less efficiency. There are allosteric activators as well as inhibitors. Allosteric activators bind to locations on an enzyme away from the active site, inducing a conformational change that increases the affinity of the enzyme's active site(s) for its substrate(s).

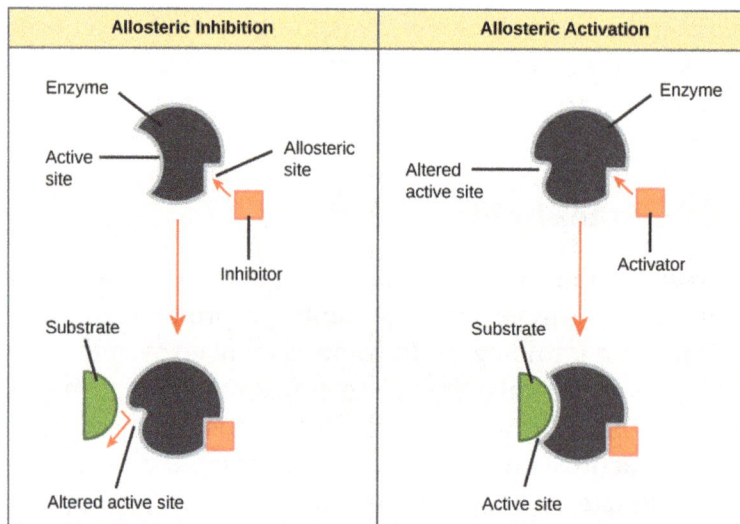

| Allosteric Inhibition | Allosteric Activation |
|---|---|

Allosteric inhibitors modify the active site of the enzyme so that substrate binding is reduced or prevented. In contrast, allosteric activators modify the active site of the enzyme so that the affinity for the substrate increases.

Many enzymes don't work optimally, or even at all, unless bound to other specific non-protein helper molecules, either temporarily through ionic or hydrogen bonds or permanently through stronger covalent bonds. These helper molecules are termed cofactors. Binding to these molecules promotes optimal conformation and function for their respective enzymes. Cofactors may be inorganic ions such as iron ($Fe^{2+}$) and magnesium ($Mg^{2+}$), and these ions maybe be linked to larger nonprotein molecules. Coenzymes are a subclass of cofactors that are organic helper molecules, with a basic atomic structure made up of carbon, nitrogen and hydrogen, which are required for enzyme action. The most common sources of coenzymes are dietary vitamins. Some vitamins are precursors to coenzymes and others act directly as coenzymes. Vitamin C is a coenzyme for multiple enzymes that take part in building the important connective tissue component, collagen. An important step in the breakdown of glucose to yield energy is catalysis by a multi-enzyme complex called pyruvate dehydrogenase. Pyruvate dehydrogenase is a complex of several enzymes that actually requires one cofactor (a magnesium ion) and five different organic coenzymes to catalyze its specific chemical reaction. Succinate dehydrogenase, an enzyme involved in both electron transport and the citric acid cycle, is another example of an enzyme that carries many cofactors, allowing it to transport electrons through the enzyme from the original donor molecule (succinate), through FAD/FADH, through a variety of other carrier, and finally to Q. Therefore, enzyme function is often made possible only via by an abundance of various cofactors.

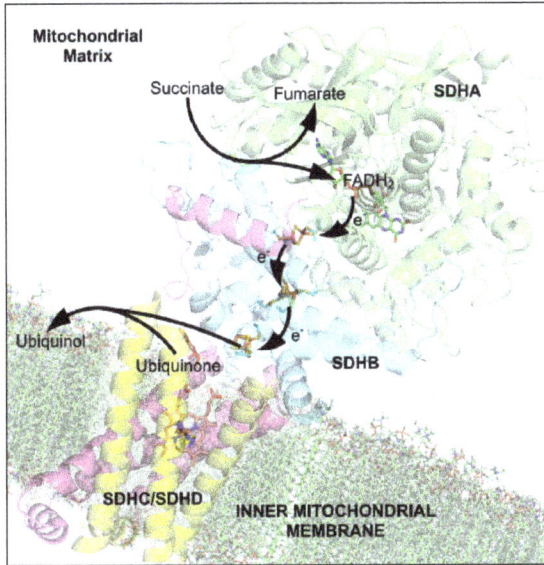

## Enzyme Compartmentalization

In eukaryotic cells, molecules such as enzymes are usually compartmentalized into different organelles. This allows for yet another level of regulation of enzyme activity. Enzymes required only for certain cellular processes can be housed separately along with their substrates, allowing for more efficient chemical reactions. Examples of this sort of enzyme regulation based on location and proximity include the enzymes involved in the latter stages of cellular respiration, which take place exclusively in the mitochondria, and the enzymes involved in the digestion of cellular debris and foreign materials, located within lysosomes.

# Factors Influencing Enzyme Activity

Enzymes are among the fastest catalysts known. It is not atypical of an enzyme to increase a reaction rates by 10,000x, 100,000x or even 1,000,000x. Indeed, some enzymes increase reaction rates to 100,000,000x the rate they would occur spontaneously. Clearly, the catalytic function of enzymes is essential to homeostasis, as without these catalysts many of the chemical processes needed for homeostasis simply would not occur quickly enough.

The ability of an enzyme to convert substrate into product is referred to as enzyme activity, and is often used as a synonym for reaction rate (since as enzyme activity increases, more substrate is converted into product per unit time). Enzyme activity is not necessarily constant—there can be a number of factors that influence how quickly substrate can be converted into product.

## Effect of Enzyme Concentration

An enzyme molecule is analogous to a worker on an assembly line in a factory. The worker picks up the raw materials, does something with them, releases the altered materials, then picks up the next set of raw materials. Likewise an enzyme molecule binds its substrate(s), catalyzes a reaction, and releases the product(s). Each step in this process requires time—time to receive the raw materials, do what needs to be done to them, and release the product. So each enzyme molecule requires x amount of time to produce one unit of product. The more enzyme molecules that are available, however, the more product can be produced in x time: two enzyme molecules would produce two units in that time period, three enzyme molecules would produce three units of product, etc. Thus the more enzyme is available, the more quickly substrate can be converted into product. In general, then, (and assuming all other factors are constant) as enzyme concentration increases, there is a proportional increase in reaction rate.

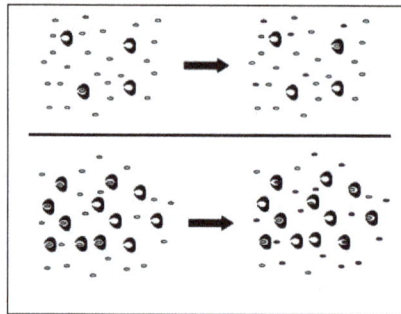

An example of the effect of enzyme concentration on reaction rate. In the top panel, four enzyme molecules are able to convert three molecules of substrate (light blue) into product (dark purple) in x time, whereas in the lower panel twelve enzyme molecules are able to produce 15 molecules of product in the same amount of time.

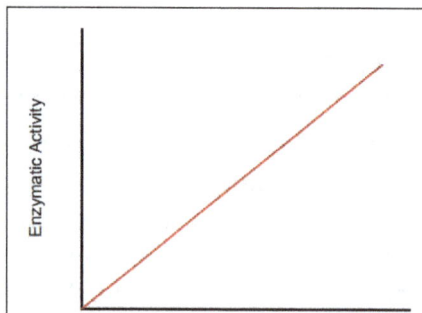

Plot illustrating the mathematical relationship between enzyme concentration and enzyme activity

## Effect of Substrate Concentration

In order for an enzyme to convert substrate into product, the substrate must first bind with the enzyme. This is usually achieved simply by random collisions between enzyme and substrate as these particles diffuse around in solution. The frequency with which these collisions occur can be influenced by a number of factors. One of them is simply the amount of each substance that is present in solution. For example, the more concentrated the substrate is within a solution, the more frequently substrate molecules will randomly collide with the enzyme in the proper orientation for the active site to bind to the substrate, and thus the more often substrate will be converted into product. So in general (again, assuming that all other factors, such as enzyme concentration, are constant), one would expect the rate of an enzyme catalyzed reaction to increase as substrate concentration increases.

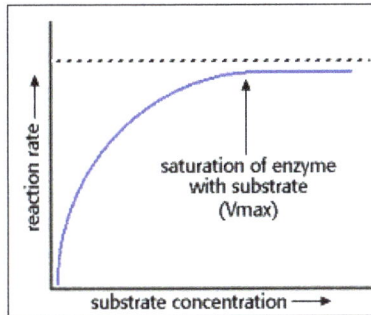

Plot illustrating the relationship between substrate concentration and reaction rate.

This is not an indefinite increase. That is because that although the rate at which substrate can bind to the enzyme increases with increasing substrate concentration, once that enzyme has bound the substrate the enzyme must still catalyze the reaction and release the product. Since there is a minimum amount of time needed to process each substrate molecule once bound, there is a maximum rate at which a given number of enzyme molecules can convert substrate into product. Thus, enzymes can demonstrate saturation in response to increasing substrate concentration. As substrate concentration increases, more and more enzyme molecules are locked up in enzyme-substrate complexes at any given time. At a particular substrate concentration the enzyme will become completely saturated with substrate—at any given time virtually all of the enzyme molecules will be occupied in enzyme-substrate complexes, and thus no further increase in reaction rate will accompany further increases in concentration.

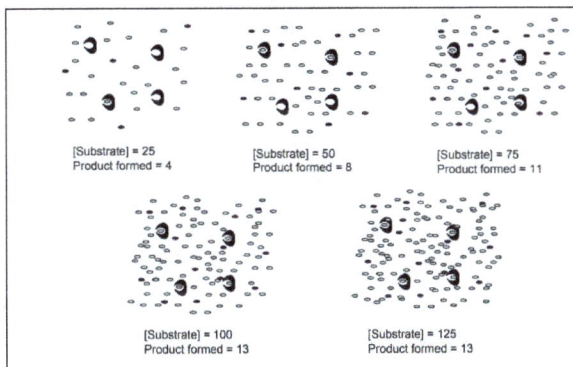

The effect of substrate concentration on reaction rate. Sets of four enzyme molecules each are exposed to different concentrations of substrate for x time. Notice that as substrate concentration increases from low levels the rate of the reaction increases, but the degree to which the rate increases become progressively less and less as more enzymes are occupied at any given time in enzyme substrate complexes. Eventually, saturation of the enzyme is achieved when all of the enzyme molecules are bound to substrate at any given time. At this point, the reaction has reached its maximum rate, and no further increase will occur with increasing substrate concentration increases, more and more enzyme molecules are locked up in enzyme-substrate complexes at any given time. At a particular substrate concentration the enzyme will become completely saturated with substrate—at any given time virtually all of the enzyme molecules will be occupied in enzyme-substrate complexes, and thus no further increase in reaction rate will accompany further increases in concentration.

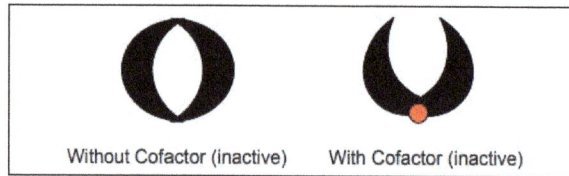

Example of a cofactor (red) used to activate an enzyme by altering the shape of its active site to the configuration needed to bond and catalyze substrates.

## Effect of Cofactor/Coenzyme Concentration

Many (but not all) enzymes require certain additional substances to be bound to them in order to function as catalysts. These substances are often referred to as cofactors and coenzymes. These auxiliary substances may need to be bound to the enzyme in order for the enzyme to have the proper shape to its active site or may be the actual catalytic agent used to facilitate the reaction taking place, whereas the enzyme merely binds the substrate and holds it in the proper orientation.

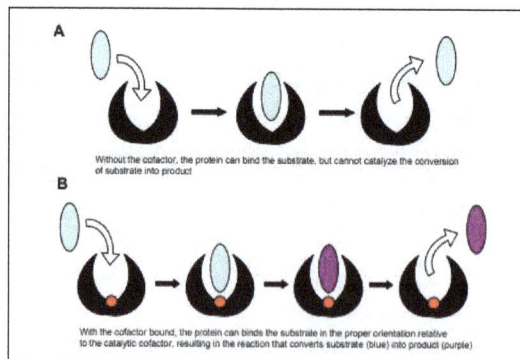

Example of an enzyme requiring a cofactor as a catalytic agent. In the top panel, the enzyme is able to bind the substrate (light blue), but without the cofactor present, it cannot convert the substrate into product. In the lower panel, the cofactor (red) acts as the catalytic agent for converting the substrate into product (dark purple).

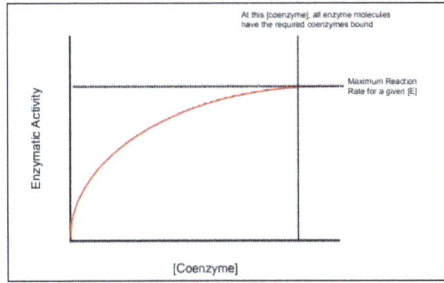

Plot of the relationship between cofactor/coenzyme concentration and reaction rate.

For those enzymes that require a cofactor or coenzyme, enzyme activity is dependent upon the concentration of that cofactor. If the cofactor is at very low concentrations, few enzyme molecules will have the necessary cofactor bound, thus few will be able to catalyze the reaction, and reaction rates will be low. As cofactor concentration increases, more and more enzyme molecules will have bound cofactor and thus be catalytically active. However, as cofactor concentration increases, there will be a progressively smaller and smaller increase in reaction rate as the majority of enzyme molecules will already have the cofactor they need. Indeed, above a certain point, virtually all enzyme molecules will have the cofactor they need, and thus increasing cofactor concentration will have no further influence on reaction rate.

## Effect of Temperature

Temperature is the average kinetic energy of a system. Kinetic energy, in turn is the energy in motion. This means that at higher temperatures particles tend to be moving more quickly than they are at slower temperatures. In solids, molecules remain in roughly the same position in space but vibrate more. In liquids and gases, where particles are free to move from one location to another, these particles tend to do so at greater speeds. Since particles are moving more quickly, they also tend to collide with one another more frequently and with greater energy. Therefore, the rates of chemical reactions (both catalyzed and non-catalyzed) tend to increase as temperature increases. Hypothetically, this should be an indefinite relationship, meaning that an increase in reaction should accompany an increase in temperature regardless of how high that temperature is Many enzymes show an unusual relationship between reaction rate and temperature.

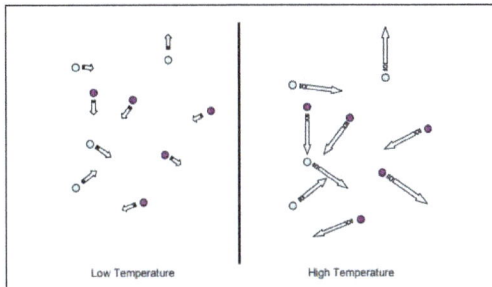

Effect of temperature on motion of particles. Increased temperature increases movement of particles in solution

Although over much of the range of temperatures biological organisms experience there is an increase in enzyme activity with increased temperature there is often a decrease in reaction rates at very high temperatures (e.g., above 70 °C). There could be a number of reasons. For example, the increase in temperature may weaken and destabilize the bonds that link enzymes with necessary cofactors, thus the rate of spontaneous deactivation of enzymes increases. However, perhaps the most important factor to consider is that the shape of the enzyme can be influenced by temperature. The secondary, tertiary, and quaternary structures of proteins all rely on relatively weak non-covalent bonds (e.g., hydrogen bonds, van der Waals forces, and ionic bonds) to link different regions of the protein together. Increasing temperature causes increased random movement in different regions of the protein, thus destabilizing these weak bonds and causing a change in the shape of the protein (denaturization). If enough of these weak bonds are broken, the shape of the active site will begin to distort, and the enzyme will lose its ability to bind substrate and catalyze the reaction. Thus the decrease in reaction rate is due to the inability of the enzyme to function as a catalyst when it is denatured by heat.

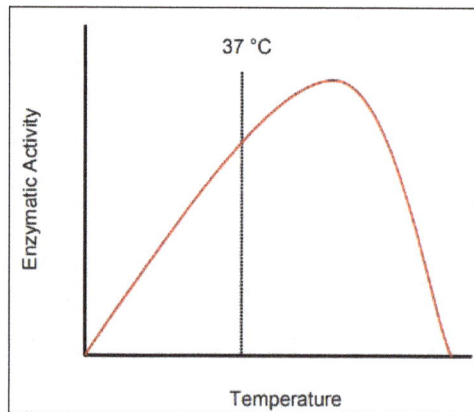

Plot illustrating the relationship between temperature and the rate of enzyme catalyzed reactions

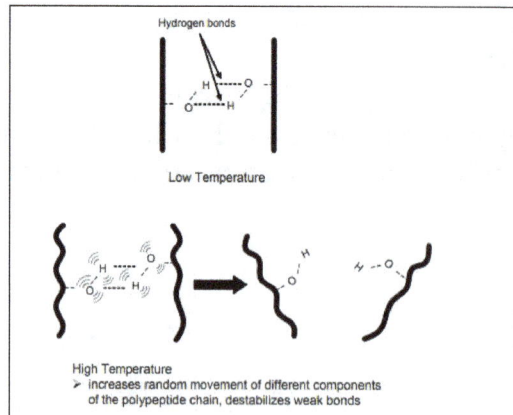

Illustration of heat denaturization between two polar amino acid side-chains. As temperature increases, random movements of the polypeptide chain pull against the hydrogen bonds linking the two side chains, causing them to destabilize and break

# Effect of pH

pH is an index of hydrogen ion ($H^+$ concentration). The $H^+$ concentration of a water-based solution can vary due to the presence of particular solutes. Some solutes, called acids, are normally weakly bound to one or more hydrogen ions, so that if dissolved in releasing one or more $H^+$ ions into solution. If the undissociated acid had a neutral charge, the water the acid tends to dissociate (break apart) dissociated acid, once losing a $H^+$ without the shared electron, would become negatively charged. Conversely, other solutes called bases lead to a reduction in $H^+$ concentrations in aqueous solutions by binding $H^+$ ions. If the base had a neutral charge before being mixed with water, it would have a positive charge upon binding a $H^+$ ion.

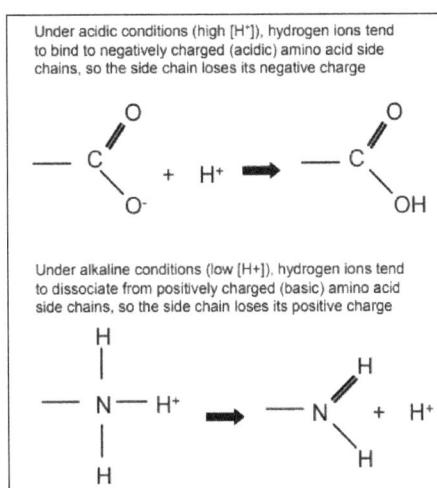

Under acidic conditions (high [$H^+$]), hydrogen ions tend to bind to negatively charged (acidic) amino acid side chains, so the side chain loses its negative charge

Under alkaline conditions (low [H+]), hydrogen ions tend to dissociate from positively charged (basic) amino acid side chains, so the side chain loses its positive charge

Examples of dissociation and binding of $H^+$ by acidic and basic amino acids under non-neutral pHs

Some amino acids have side chains that can act as weak acids or weak bases. At a pH of 7, acidic side chains will tend to be dissociated (will have a negative charge) whereas basic side chains will have bound a $H^+$ ion (will have a net positive charge). Since these side chains have net charges, they can form ionic interactions with one another, where like charged side chains repel one another and opposite charged side chains attract one another. These types of interactions contribute to the tertiary and quaternary structure of proteins. tendency for an acid or a base to be bound to a $H^+$ or to release a $H^+$ is influenced largely by [$H^+$] in the surrounding environment. If the [$H^+$] is high (low pH), then the side chain will tend to Fig. Examples of dissociation and binding of H+ by acidic and basic amino acids under non-neutral pHs. be bound to $H^+$ so acidic side chains would be neutrally charged and basic side chains would be positively charged. Conversely, if [$H^+$] is low, then side chains will tend to release a bound H+, so the acidic side chains would be negatively charged, and the basic side chains neutrally charged. Note that this would dramatically alter the ionic interactions that could otherwise occur among acidic and basic amino acids—ionic bonds that would exist at pH 7 would no longer exist at higher or lower pHs. This in turn, could radically alter the tertiary and quaternary structure of the protein. Since the catalytic ability of an enzyme is so tightly linked to the specific shape and chemical properties of its active site, alteration of normal ionic bonding patterns within the protein tends to reduce catalytic function. For any enzyme, then, there is an optimal pH where the right degree of $H^+$ binding and dissociation of various acidic and basic amino acids exists such that the active site of the protein has the shape for maximum catalytic activity. Deviations of pH from this optimal level, to lesser or greater degrees, tend to reduce the ability of the enzyme to catalyze its reaction. The optimal pH can vary considerably among enzymes—whereas most enzymes in the human body function optimally at a pH that is roughly neutral (cytoplasm and most extracellular fluids tend to have a pH between 7 and 8), some enzymes, such as those of the digestive system, may have optimal pHs at very acidic or very alkaline levels.

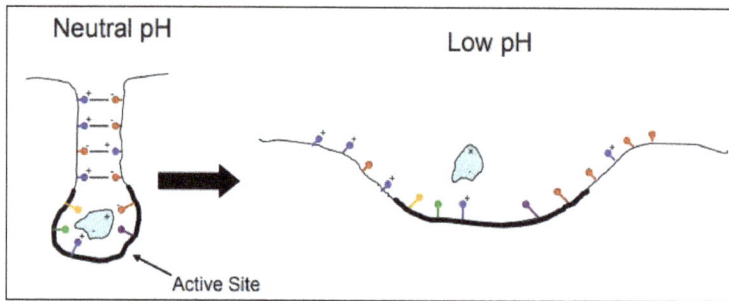

Denaturization of an enzyme when exposed to acidic conditions. Note that because of the high $[H^+]$ both basic (blue) and acidic (red) side chains are bound to $H^+$ when exposed to low pH. As a result, the ionic bonds that normally would hold the shape of the active site in its proper confirmation at neutral pH (left) cannot form at low pH (right). The enzyme thus is unable to bind substrate and catalyze its conversion into product.

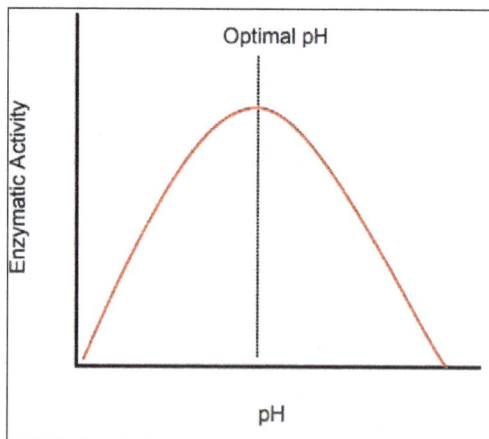

Plot illustrating variation in the rate of enzyme-catalyzed reactions with pH.

# References

- Enzyme-Kinetics, life-sciences: news-medical.net, Retrieved 22 April, 2019

- Enzyme-inhibitors-and-activators: intechopen.com, Retrieved 9 February, 2019

- Enzyme-purification: creative-enzymes.com, Retrieved 11 August, 2019

- Enzyme-Purification-By-Electrophoresis: creative-enzymes.com, Retrieved 23 January, 2019

- Michaelis-Menten-Kinetics, Enzymatic-Kinetics, Enzymes, Biological-Chemistry: libretexts.org, Retrieved 19 May, 2019

- Enzyme-substrate-complex: biologydictionary.net, Retrieved 25 March, 2019

# 3

# Active Site and Enzyme Cofactors

The molecules on which the enzymes act are known as substrates. The region where substrate molecules bind and undergo a chemical reaction is known as an active site. In order for the enzyme to function as a catalyst, a non-protein chemical compound or a metallic ion known as an enzyme cofactor is required. This chapter has been carefully written to provide an easy understanding of the varied facets of active sites and enzyme cofactors.

## Active Site

The active site of an enzyme is the site which shows the highest metabolic activity by catalysing the enzyme-substrate complex into the products. The active site is found deep inside the enzyme which resembles a hole or small depression. An active site is a region which attaches the substrate molecule with the enzyme and thus catalysing the reaction.

As we know the enzyme is "Highly specific" molecule, but its specificity is due to the active site which allows the binding of a particular substrate. The amino acids residues are present around the active site which holds the substrate molecule at the right position while the reaction takes place. The substrate molecule shows high binding affinity towards the active site.

The active site can define as the small region which appears like a cleft or cavity which is composed of about 10-15 amino acid residues. it is a site which activates the complex enzyme to bind with the particular substrate and induces the transition state of the substrate and stabilize the product formation. The active site also refers as Catalytic site.

The active site performs two functional activities:

1. Binding activity

2. Catalytic activity

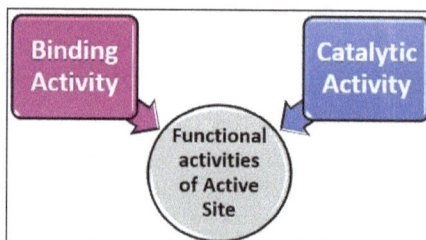

- Binding Activity: The binding activity is a property of active site which increases the binding affinity of the substrate with an enzyme.

- Catalytic activity: It is a property of an active site which carries out the catabolic reaction where the enzyme and substrate react to form a product by reducing the activation energy.

## Reaction Mechanism of an Enzyme

In an enzyme-catalyzed reaction, the substrate will attach to the active site of an enzyme. A specific substrate will bind to the active site of an enzyme. After the attachment of substrate and enzyme, an "Enzyme-substrate complex" forms. In the E-S complex, the substrate on enzyme activity will convert into a product. At last, the product gets released and the enzyme becomes free to reuse again.

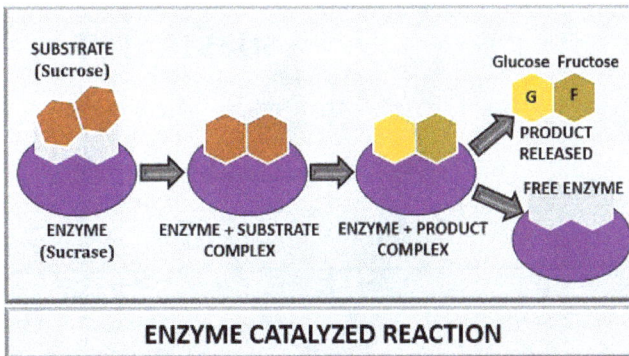

**ENZYME CATALYZED REACTION**

Example:

Let us take an example, where sucrose is a substrate which combines with the active site of an enzyme "Sucrase". After the binding of sucrose with an enzyme Sucrase, an E-S complex forms. Then a reaction between Sucrase and Sucrose takes place. The reaction will change the structural conformation of the Sucrose refers to as "Transition state of the Sucrose". The change in the structural configuration of a Sucrose leads to the conversion of the E-S complex into the E-P complex. At last, glucose and fructose released as products form the Sucrase enzyme.

## Characteristics

- An active site is a specific location found in the enzyme where a substrate binds to catalyze the reaction. It refers as "Enzyme catalytic surface".

- About 10-15 amino acid residues combine to form an active site.

- The active site possesses a specific geometrical shape and chemical signals which allow the enzyme for the recognition and binding of the specific substrate.

- An active site will allow the specific substrate to bind whose shape is complementary to the active site. Therefore, a substrate is like is a key which can only fit into the particular lock i.e. active site.

- The active site of an enzyme catalyzes many chemical or biological pathways.

- After the formation of the enzyme-substrate complex, both substrate and active site change its structural configuration by bending the target bonds and breaking the substrate molecule into a product.

- Enzymes show "Catalytic activity" which is due to its active site. It catalyzes a substrate into a product after complementary binding of the substrate with the active site of an enzyme based on geometric shape, size, charge and stereospecificity etc.

- The active site of an enzyme induces the "Transition of the substrate".

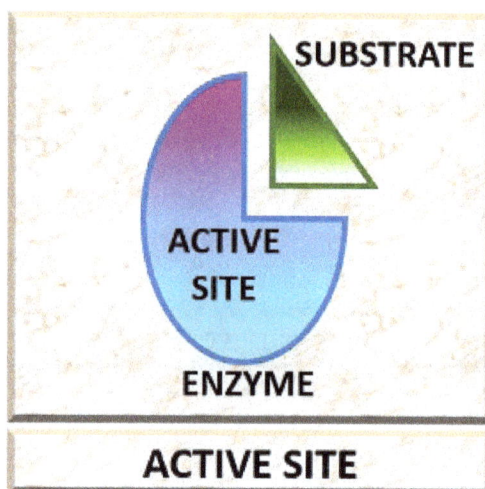

ACTIVE SITE

## Properties

There are following characteristics of an active site which includes:

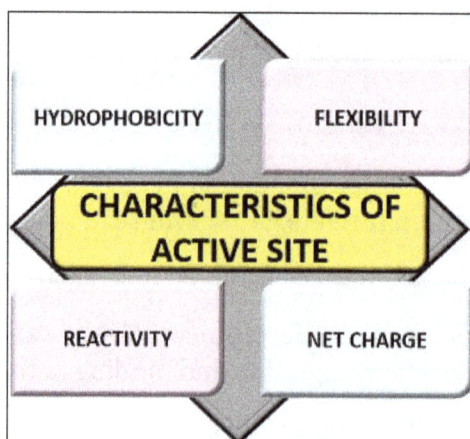

## Hydrophobicity

The initial binding of substrate and enzyme is through the non-covalent bond. But the catalytic site involves hydrophobic interaction for the attachment of the substrate with the enzyme. Hydrophobic binding of the substrate to the active site of an enzyme increases the binding affinity. Other than hydrophobic interaction, there are three other mechanisms also like Vander Waal, hydrogen bond and electrostatic force of interaction which promotes the formation of E-S complex.

## Flexibility

An active site shows flexibility as it can change its conformation which catalyzes the conversion of the substrate into a product.

## Reactivity

The active site of an enzyme reacts with the specific substrate. Its reactivity depends upon the environmental conditions like temperature, pH, the concentration of enzyme and substrate etc. The active site of an enzyme combines with the substrate and lowers the activation energy to catalyze the reaction.

## Net Charge

The active site mainly consists of non-polar amino acid residues which carry no charge or having a o net charge. Some active site also consists of polar amino acids which can carry both positive and negative charge. The net charge of the catalytic site decides which amino acid will bind with the enzyme. There must be a complementary pairing between the active site and the substrate. The same charge on both the catalytic site and substrate will not form an E-S complex as there will occur repulsion between the two.

## Role of Active Site

As we have discussed the active site performs two major activities like:

- The binding of a substrate with an enzyme.
- The catalytic activity by conversion of substrate into product.

Let us assume two conditions, One is the conversion of substrate into a product without enzyme and second in the presence of an enzyme.

## First Condition

In the first condition, we will discuss the transition reaction of the substrate into a product in the absence of an enzyme catalyst. For this, plot a graph between Reaction direction and Energy. In the absence of a "Catalyst", a substrate (S) will require higher

"Activation energy" to go into the transition state which we will represent as "St". In the transition state, the substrate will change its conformation and thereby release a product (P).

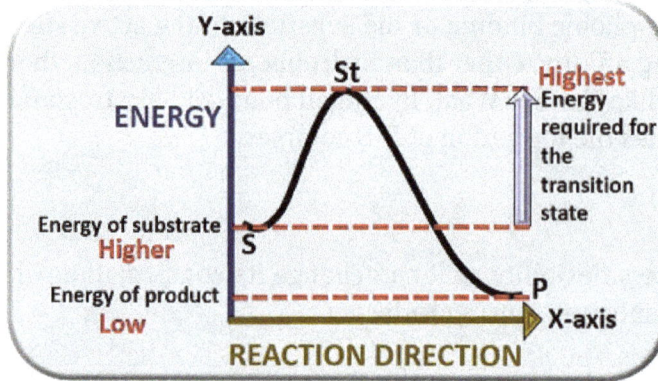

## Second Condition

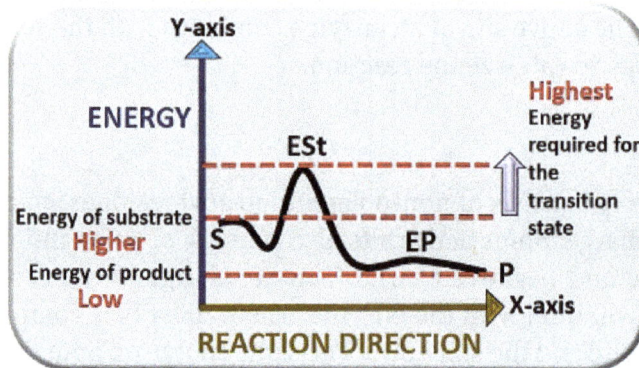

Here, we will discuss the transition reaction of the substrate into a product in the presence of an enzyme catalyst. For this also, plot a graph between Reaction direction and Energy. In the presence of a "Catalyst", a substrate (S) will bind to the catalytic site of an enzyme. The enzyme being a catalytic agent undergo catalysis of a substrate. The enzyme will modify the substrate and takes it to the transition state which we will represent as "ESt". In the transition state the enzyme and substrate will react and there a change occurs in the configuration of the substrate. This change will lead to the formation of the enzyme product complex and finally release a product (P).

Therefore, the active site of an enzyme lowers the activation energy by increasing the rate of reaction. The energy level of the substrate is higher than that of a product but lower than the transition state of the substrate. The activation energy is inversely proportional to the rate of reaction thus it decreases with the increase in the rate of reaction.

# Cofactor

Cofactor is a non-protein chemical that assists with a biological chemical reaction. Co-factors may be metal ions, organic compounds, or other chemicals that have helpful properties not usually found in amino acids. Some cofactors can be made inside the body, such as ATP, while others must be consumed in food.

Minerals, for example, come from the environment, and cannot be made from scratch by any living cell. The organic compounds we refer to as "vitamins" are cofactors that our own bodies cannot make, so we must consume them from food in order for our cells to be able to perform essential life functions.

At the biochemical level, cofactors are important in understanding how biological reactions proceed. The presence or absence of cofactors may determine how quickly reactions proceed from their reactant to their product.

At the biological level, understanding cofactors is important to understanding health. Without the proper cofactors, humans and other animals can develop serious diseases and even death.

## Function of Cofactors

Cofactors generally serve the purpose of supplying chemical groups or properties that are not found in other chemical groups.

ATP, for example, is a cofactor with a unique ability to transfer energy to drive chemical processes such as the activity of enzymes and transport proteins.

There are dozens of known cofactors, each of which may be necessary for multiple biochemical reactions, as illustrated below.

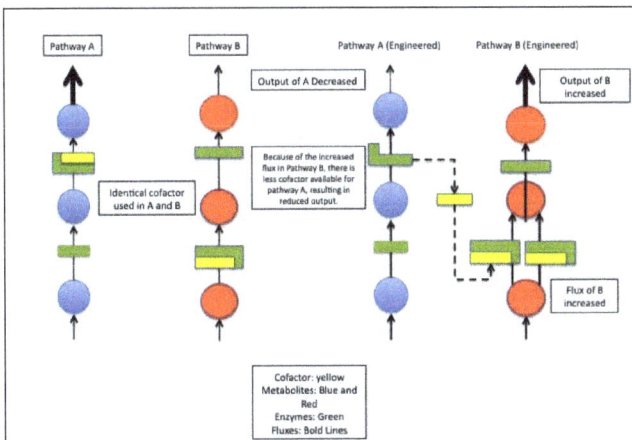

As a result, the functions of cofactors may be as diverse as their chemical structures and properties.

The wide-ranging effects of cofactors can be seen by studying vitamin deficiencies: deficiencies of different vitamins, many of which are cofactors, have many different negative effects on human health.

## Types of Cofactor

## Vitamins

Vitamins are organic compounds that are co-factors for necessary biochemical reactions. Vitamins typically need to be consumed in the diet, because they cannot be made inside the body.

Many vitamins are cofactors which help enzymes to catalyze reactions, such as the production of important proteins. Vitamin C, for example, is a cofactor for the production of the connective tissue collagen.

This is why people who get scurvy – a severe form of vitamin C deficiency – may experience connective tissue problems, including muscle weakness, muscle soreness, and even unexplained bleeding as the connective tissues of blood vessels cannot be replaced.

Vitamin deficiencies are a good illustration of the effects of co-factor deficiency. Just as there are many possible vitamin deficiencies with many different symptoms, there are many different co-factors that our body needs to carry out its diverse necessary biochemical reactions.

The body's requirement for diverse vitamin cofactors is also the reason why nutritionists counsel people to "eat the rainbow" – many plants' colors are produced by cofactors, so by eating fruits and vegetables in a wide variety of colors helps to ensure that we consume a healthy variety of cofactors.

## Minerals

Like vitamins, minerals are chemicals from outside of the body that must be ingested to allow our cells to function properly. The difference is that while vitamins are organic molecules – molecules containing carbon, which are often made by other living things – minerals are inorganic substances that occur naturally, and are often found in rocks and soil.

Minerals often enter our diets from plants, which draw them up out of the ground through their roots along with water. In some rare cases, people with vitamin deficiencies may feel the urge to eat certain types of soil to obtain the minerals from the soil directly.

Minerals that are important for human health include copper, which is necessary for the function of some important liver enzymes that break down toxins; iron, which is necessary for the function of some important metabolic enzymes; magnesium, which is necessary for the function of DNA polymerase and other enzymes; and zinc, which is also necessary for DNA polymerase as well as some liver enzymes.

As with vitamins, there can be too much of a good thing – while minerals are necessary in small amounts for our metabolisms to function, taking large doses of them can result in toxicity and death. Indeed, overdoses of iron-containing multivitamins are a leading cause of death in children under 4, who may mistake these multivitamins for candy.

## Organic Non-vitamin Cofactors

Some cofactors are organic substances not classified as enzymes. Some of these may be made inside our own bodies, and so not qualified as vitamins.

Organic non-vitamin cofactors include ATP – an essential assistant to many biochemical processes, which transfers energy to numerous enzymes, transport proteins, and more; coenzyme Q, which plays a vital role in the mitochondrial transport chain; and heme, which is a complex iron-containing compound that is necessary for our blood cells to carry oxygen throughout our bodies.

## Examples of Cofactors

### Thiamine (Vitamin B3)

Thiamine is a vitamin found primarily in edible seeds such as beans, corn, and rice. To improve public health, thiamine is often artificially added to wheat-containing products such as breakfast cereals.

In the body, thiamine is used to make many co-enzymes that assist with important processes. It is made into thiamine pyrophosphate, which is necessary to break down sugars and amino acids.

Severe thiamine deficiency is one cause of Korsakoff Syndrome – a rare neurological disorder seen in people with severe alcohol addiction. In Korsakoff Syndrome, severe malnutrition, lack of thiamine, and brain damage from overuse of alcohol combine to produce severe symptoms, including memory impairment. Some sufferers of Koraskoff Syndrome are not able to form new memories because the metabolism of their brain is so impaired.

### Folic Acid (Vitamin B9)

Folic acid is another vitamin which is now often added to food to improve public health. It is necessary for the body to produce DNA, RNA, and amino acids, which are necessary for growth and cell division.

This makes folic acid particularly essential for pregnant women, whose fetuses are producing new cells and tissues very quickly. Deficiencies in folic acid can lead to birth defects in babies, or to anemia in pregnant women who may not be able to make enough new blood cells to supply both them and the baby.

For this reason, it is recommended that all women of childbearing age talk to their doctors about taking folic acid supplements. Pregnancy outcomes are best when sufficient folic acid is present in the mother's body even before pregnancy begins.

## Iron-sulfur Clusters

Iron-sulfur clusters are clusters of iron and sulfur ions which can form stable arrangements. These clusters have many unique properties that are not found in amino acids or other organic compounds.

The unique properties of iron-sulfur clusters make them very useful for biological reactions involving electron transfers. Both iron and sulfur are able to store and release electrons with greater eases than more common atoms such as carbon.

This makes iron-sulfur clusters a vital part of cofactors and enzymes involved in electron transfer and energy transfer, including NADH dehydrogenase, coenzyme Q, cytochrome C, and Complex I and Complex II in the mitochondria.

# Coenzyme

A coenzyme is an organic non-protein compound that binds with an enzyme to catalyze a reaction. Coenzymes are often broadly called cofactors, but they are chemically different. A coenzyme cannot function alone, but can be reused several times when paired with an enzyme.

## Functions of Coenzymes

An enzyme without a coenzyme is called an apoenzyme. Without coenzymes or cofactors, enzymes cannot catalyze reactions effectively. In fact, the enzyme may not function at all. If reactions cannot occur at the normal catalyzed rate, then an organism will have difficulty sustaining life.

When an enzyme gains a coenzyme, it then becomes a holoenzyme, or active enzyme. Active enzymes change substrates into the products an organism needs to carry out essential functions, whether chemical or physiological. Coenzymes, like enzymes, can be reused and recycled without changing reaction rate or effectiveness. They attach to a portion of the active site on an enzyme, which enables the catalyzed reaction to occur. When an enzyme is denatured by extreme temperature or pH, the coenzyme can no longer attach to the active site.

# Examples of Coenzymes

Most organisms cannot produce coenzymes naturally in large enough quantities to be effective. Instead, they are introduced to an organism in two ways:

## Vitamins

Many coenzymes, though not all, are vitamins or derived from vitamins. If vitamin intake is too low, then an organism will not have the coenzymes needed to catalyze reactions. Water-soluble vitamins, which include all B complex vitamins and vitamin C, lead to the production of coenzymes. Two of the most important and widespread vitamin-derived coenzymes are nicotinamide adenine dinucleotide (NAD) and coenzyme A.

NAD is derived from vitamin B3 and functions as one of the most important coenzymes in a cell when turned into its two alternate forms. When NAD loses an electron, the low energy coenzyme called $NAD^+$ is formed. When NAD gains an electron, a high-energy coenzyme called NADH is formed.

$NAD^+$ primarily transfers electrons needed for redox reactions, especially those involved in parts of the citric acid cycle (TAC). TAC results in other coenzymes, such as ATP. If an organism has a $NAD^+$ deficiency, then mitochondria become less functional and provide less energy for cell functions.

When $NAD^+$ gains electrons through a redox reaction, NADH is formed. NADH, often called coenzyme 1, has numerous functions. In fact, it is considered the number one coenzyme in the human body because it is necessary for so many different things. This coenzyme primarily carries electrons for reactions and produces energy from food. For example, the electron transport chain can only begin with the delivery of electrons from NADH. A lack of NADH causes energy deficits in cells, resulting in widespread fatigue. Additionally, this coenzyme is recognized as the most powerful biological antioxidant for protecting cells against harmful or damaging substances.

Coenzyme A, also known as acetyl-CoA, naturally derives from vitamin B5. This coenzyme has several different functions. First, it is responsible for initiating fatty acid production within cells. Fatty acids form the phospholipid bilayer that comprises the cell membrane, a feature necessary for life. Coenzyme A also initiates the citric acid cycle, resulting in the production of ATP.

## Non-Vitamins

Non-vitamin coenzymes typically aid in chemical transfer for enzymes. They ensure physiological functions, like blood clotting and metabolism, occur in an organism. These coenzymes can be produced from nucleotides such as adenosine, uracil, guanine, or inosine.

Adenosine triphosphate (ATP) is an example of an essential non-vitamin coenzyme. In fact, it is the most widely distributed coenzyme in the human body. It transports substances and supplies energy needed for necessary chemical reactions and muscle contraction. To do this, ATP carries both a phosphate and energy to various locations within a cell. When the phosphate is removed, the energy is also released. This process is result of the electron transport chain. Without the coenzyme ATP, there would be little energy available at the cellular level and normal life functions could not occur.

Here is an example of the electron transport chain. The vitamin-derived coenzyme NADH begins the process by delivering electrons. ATP is the final resulting product:

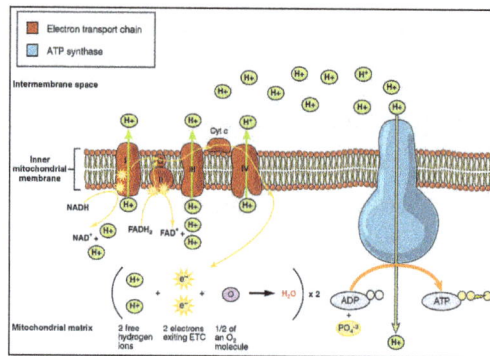

## Coenzyme A

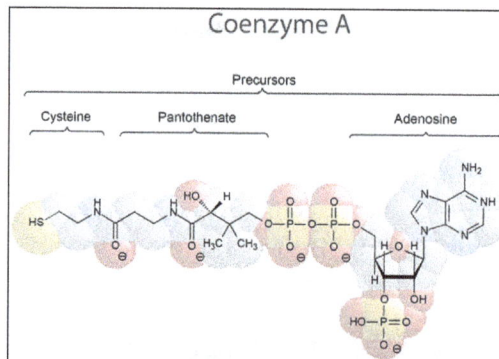

Coenzyme A (CoA, CoASH or HSCoA) is the key cofactor in first step of the TCA cycle, responsible for transferring the acetyl group from pyruvate oxidation to oxaloacetate yielding citrate.

Coenzyme A is also a critical cofactor in fatty acid metabolism. Coenzyme A carries fatty acids through the catabolic/oxidation process in the mitochondria and transfers acetyl groups during the elongation process of fatty acid synthesis in the cytosol.

The acetyl moiety of acetyl CoA is bound by a high-energy bond (free energy 34.3 kJ/mol) to the -SH group of Coenzyme A. It is also a precursor to, steroids and other naturally occurring compounds, such as terpenes and acetogenins present in plants.

In the transfer reaction by Acetyl CoA of the C2 acetyl fragment, either the carboxyl group or the methyl group may react (electrophilic vs. nucleophilic reaction, respectively).

AcetylCoA is prepared enzymatically by reacting Coenzyme A with Acetyl Phosphate and Phosphotransacetylase. The product is purified by ion exchange chromatography. Several methods of preparation and methods for the determination of Acetyl CoA and other CoA derivatives have been described in the literature. Coenzyme A is synthesized in vivo from pantothenate, cysteine, and adenosine. Pantothenate is phosphorylated, joined with cysteine, decarboxylated, joined with adenosine and phosphorylated again to the 3' ribose position to yield Coenzyme A.

## The Functions of Coenzyme A

### Fatty Acid Synthesis

Coenzyme A is the helper molecule that facilitates the oxidation pathway. This process results in the production of acetyl-coenzyme A, an important chemical substance used for the initiation of fatty acid production within the living cell. Without this much-needed process, there is no production of fatty acids, the compounds that maintain the integrity of the cell membrane, the protective covering of each living cell.

### Drug and Enzyme Functioning

Coenzyme A improves the functioning of some proteins, sugars and drugs, In drugs, it is used to extend a medication's half-life, the length of time needed to decay or inactivate half of active ingredients of a certain drug, prolonging its ideal effect in the body. In cells, coenzyme A causes activation or inactivation of certain chemical compounds, such as enzymes.

### Energy Production

Coenzyme A, in the form of acetyl-coenzyme A, initiates the Krebs cycle, a chemical process within the body that results in the production of carbon dioxide and adenosine triphosphate. ATP is an important, energy-rich compound that provides fuel and energy needed for the synthesis of protein and deoxyribonucleic acid, the genetic code needed for cell replication in the body.

### Coenzyme A Biosynthesis and Degradation in Eukaryotic Cells

Coenzyme A (CoA) is a fundamental cofactor in all living organisms. It has a unique chemical structure which allows the diversity in biochemical reaction products and regulatory mechanisms. A classical pathway for CoA biosynthesis involves five enzymatic steps that are highly conserved from prokaryotes to eukaryotes and utilise pantothenate (vitamin B5), adenosine triphosphate (ATP) and cysteine. The pathway is initiated by pantothenate kinase (PANK), which converts pantothenate into 4′-phosphopantothenate. 4′-Phosphopantothenoylcysteine synthase (PPCS) and phosphopantothenoylcysteine decarboxylase (PPCDC) catalyse the formation of 4′-phosphopantothenoylcysteine and 4′-phosphopantetheine (4′-PP), respectively. The last two steps in the CoA biosynthetic pathway are catalysed by CoA synthase (CoASy), which possesses two enzymatic activities: 4′-PP adenyltransferase (PPAT) and dephospho-CoA kinase (DPCK). An alternative route for CoA biosynthesis has been recently uncovered under conditions when the conventional de novo pathway is impaired and the level of intracellular CoA is significantly reduced. It has been proposed that intracellular CoA pools could be replenished through the degradation of external sources of CoA (diet or culture medium) by ectonucleotide pyrophosphatases (ENPPs) to 4′-phosphopantetheine (P-PanSH), which is then transported into a cell and incorporated in the CoA biosynthetic pathway downstream of PPCDC. The proposed mechanism requires further validation, especially the existence of dedicated P-PantSH transporters on cell membranes.

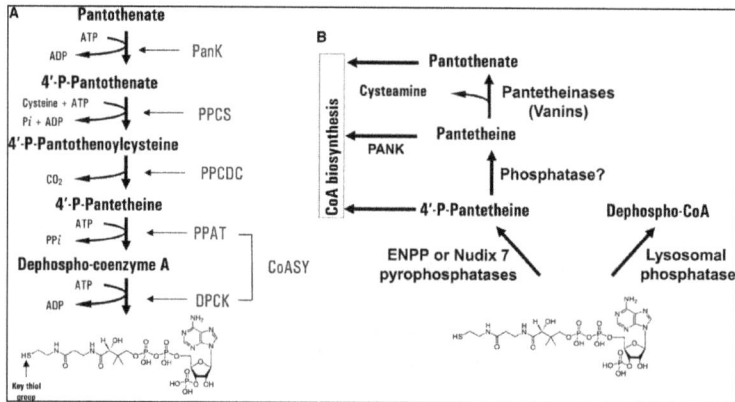

Biosynthesis and degradation of CoA in mammalian cells (A) The conventional de novo and alternative pathways of CoA biosynthesis are shown. (B) CoA degradation involves phosphodiesterases, phosphatases and pantetheinases

The biosynthesis and homeostasis of CoA is controlled at different levels: transcription of genes encoding biosynthetic enzymes, regulation of enzymatic activities by a feedback mechanism, signalling pathways, degradation of CoA and interconversion between CoA and its thioester derivatives. Various extracellular stimuli, such as nutrients, hormones of metabolic homeostasis, intracellular metabolites and stress, were found to regulate the total level of CoA in mammalian cells. It is reduced in to insulin, glucose, pyruvate and fatty acids, whereas glucagon, glucocorticoids and oxidative stress have an opposite effect. PANK is the master regulator of CoA biosynthesis. There are four PANK isoforms in mammals, which exhibit a different pattern of expression, subcellular localisation and mode of regulation, allowing them to sense and control levels of CoA/CoA derivatives in various cellular compartments. The expression and activity of the PANK proteins are governed by multiple mechanisms. Feedback inhibition by CoA/CoA thioesters (primarily acetyl CoA) is the principal mechanism for controlling the activity of mammalian PANK.

The PPAT and DPCK activities of CoA synthase were found to be strongly induced by phospholipids. The identification of CoA synthase in signalling complexes with ribosomal S6 kinase (S6K), class 1A phosphatidylinositol 3-kinase (PI3K), Src family kinases and enhancer of mRNA-decapping protein 4 (EDC4) suggests the regulation of CoA biosynthesis via signal transduction pathways and stress response.

The total cellular CoA content is also controlled by degradation, involving phosphodiesterases, phosphatases and pantetheinases (Figure). The degradation of CoA results in the generation of products which are known intermediates in the biosynthetic pathway. CoA was found to be dephosphorylated at the 3′ position of the ribose ring by a lysosomal alkaline phosphatase, leading to the formation of dephospho-CoA. Several peroxisomal and mitochondrial nucleotide diphosphate hydrolases (Nudix) were shown to hydrolyse CoA and acyl CoA thioesters to yield 3′,5′-adenosine mononucleotide and 4′-phosphopantetheine or acyl-phosphopantetheine. The degradation of extracellular CoA was found to be mediated by ENPP, which functions as a phosphodiesterase and produces 3′,5′-ADP and 4′-phosphopantetheine. Dephosphorylation of 4′-phosphopantetheine by phosphatases produces pantetheine, which is further degraded to pantothenate and cysteamine by pantetheinases (Figure). Produced pantothenate may re-enter the CoA biosynthetic pathway or be excreted.

## CoA Content and Subcellular Localisation

The estimated CoA levels in mammalian cells and tissues span more than a 10-fold range. Liver, heart and brown adipose tissue have the highest CoA levels, followed by kidney and brain. The CoA pool is largely made up of CoASH, and acetyl CoA is the largest component of the acyl CoA pool. The subcellular distribution of CoA in mammalian cells reflects the variety of processes in which it is implicated. The concentration of CoA in mitochondria and peroxisomes are in the range of 2–5 and 0.7 mM, respectively, whereas levels of cytosolic and nuclear CoA are significantly lower, ranging from 0.05 to 0.14 mM. CoA is a large and charged molecule, therefore, it must be distributed to subcellular organelles via dedicated transporters. High-affinity transporters for CoA and dephospho-CoA were identified on mitochondrial and peroxisomal membranes.

## Cellular Functions of CoA and its Thioester Derivatives

CoA and its thioester derivatives play important roles in numerous biosynthetic and degradative pathways of cellular metabolism, allosteric interactions and the regulation of gene expression. These include synthesis and oxidation of fatty acids, the Krebs cycle, ketogenesis, biosynthesis of cholesterol and acetylcholine, degradation of amino acids, regulation of gene expression and cellular metabolism via protein acetylation and others (Figure). Abnormal biosynthesis and homeostasis of CoA and its derivatives are associated with various human pathologies, including diabetes, Reye's syndrome, cancer, vitamin B12 deficiency and cardiac hypertrophy. Genetic studies in human and animal models revealed the importance of the CoA biosynthetic pathway for the development and functioning of the nervous system. Mutations in the human PANK2 and COASY genes were found to be associated with a degenerative brain disorder, termed neurodegeneration with brain iron accumulation (NBIA).

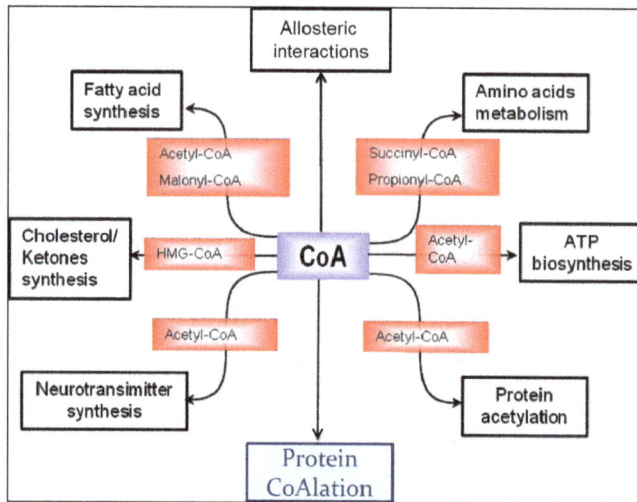

Cellular functions of CoA and its derivatives.

CoA thioester derivatives are implicated in diverse cellular functions, including the Krebs cycle, ketogenesis, biosynthesis of cholesterol and acetylcholine, the degradation of amino acids, the synthesis and oxidation of fatty acids, biosynthesis of neurotransmitters and the regulation of gene expression. Protein CoAlation is a novel, unconventional function of CoA in redox regulation and antioxidant defence.

Although de novo CoA biosynthesis is an evolutionary conserved biochemical process, significant structural and regulatory differences between microbial and human biosynthetic enzymes make the CoA biosynthetic pathway an attractive target for the development of novel antibiotics.

## Flavin Adenine Dinucleotide

In biochemistry, flavin adenine dinucleotide (FAD) is a redox-active coenzyme associated with various proteins, which is involved with several important enzymatic reactions in metabolism. A flavoprotein is a protein that contains a flavin group, this may be in the form of FAD or flavin mononucleotide (FMN). There are many flavoproteins besides components of the succinate dehydrogenase complex, including α-ketoglutarate dehydrogenase and a component of the pyruvate dehydrogenase complex.

FAD can exist in four different redox states, which are the flavin-N(5)-oxide, quinone, semiquinone, and hydroquinone. FAD is converted between these states by accepting or donating electrons. FAD, in its fully oxidized form, or quinone form, accepts two electrons and two protons to become $FADH_2$ (hydroquinone form). The semiquinone ($FADH^{\cdot}$) can be formed by either reduction of FAD or oxidation of $FADH_2$ by accepting or donating one electron and one proton, respectively. Some proteins, however, generate and maintain a superoxidized form of the flavin cofactor, the flavin-N(5)-oxide.

## Properties

Flavin adenine dinucleotide consists of two main portions: an adenine nucleotide (adenosine monophosphate) and a flavin mononucleotide bridged together through their phosphate groups. Adenine is bound to a cyclic ribose at the 1' carbon, while phosphate is bound to the ribose at the 5' carbon to form the adenine nucledotide. Riboflavin is formed by a carbon-nitrogen (C-N) bond between a isoalloxazine and a ribitol. The phosphate group is then bound to the on the terminal ribose carbon to form a FMN. Because the bond between the isoalloxazine and the ribitol is not considered to be a glycosidic bond, the flavin mononucleotide is not truly a nucleotide. This makes the dinucleotide name misleading; however, the flavin mononucleotide group is still very close to a nucleotide in its structure and chemical properties.

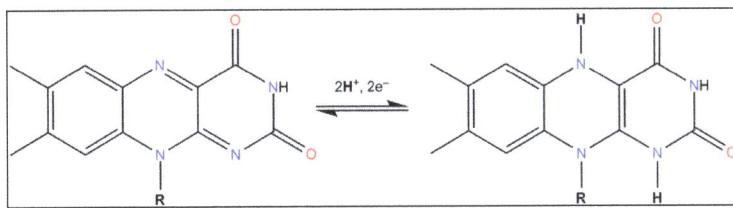

Reaction of FAD to form $FADH_2$

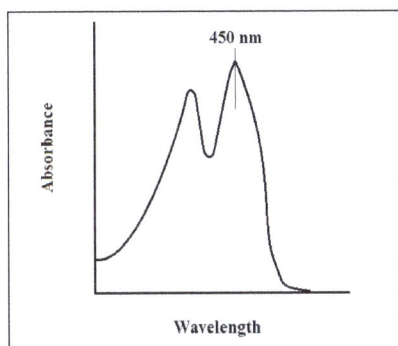

Approximate absorption spectrum for FAD

FAD can be reduced to $FADH_2$ through the addition of 2 $H^+$ and 2 $e^-$. $FADH_2$ can also be oxidized by the loss of 1 $H^+$ and 1 $e^-$ to form FADH. The FAD form can be recreated through the further loss of 1 $H^+$ and 1 $e^-$. FAD formation can also occur through the reduction and dehydration of flavin-N(5)-oxide. Based on the oxidation state, flavins take specific colors when in aqueous solution. flavin-N(5)-oxide (superoxidized) is yellow-orange, FAD (fully oxidized) is yellow, FADH (half reduced) is either blue or red based on the pH, and the fully reduced form is colorless. Changing the form can have a large impact on other chemical properties. For example, FAD, the fully oxidized form is subject to nucleophilic attack, the fully reduced form, $FADH_2$ has high polarizability, while the half reduced form is unstable in aqueous solution. FAD is an aromatic ring system, whereas FADH$_2$ is not. This means that FADH$_2$ is significantly higher in energy, without the stabilization through resonance that the aromatic structure provides. FADH$_2$ is an energy-carrying molecule, because, once oxidized it regains aromaticity and releases the energy represented by this stabilization.

The spectroscopic properties of FAD and its variants allows for reaction monitoring by use of UV-VIS absorption and fluorescence spectroscopies. Each of the different forms of FAD have distinct absorbance spectra, making for easy observation of changes in oxidation state. A major local absorbance maximum for FAD is observed at 450 nm, with an extinction coefficient of 11,300. Flavins in general have fluorescent activity when unbound (proteins bound to flavin nucleic acid derivatives are called flavoproteins). This property can be utilized when examining protein binding, observing loss of fluorescent activity when put into the bound state. Oxidized flavins have high absorbances of about 450 nm, and fluoresce at about 515-520 nm.

## Chemical States

In biological systems, FAD acts as an acceptor of H$^-$ and e$^-$ in its fully oxidized form, an acceptor or donor in the FADH form, and a donor in the reduced FADH$_2$ form. The diagram below summarizes the potential changes that it can undergo.

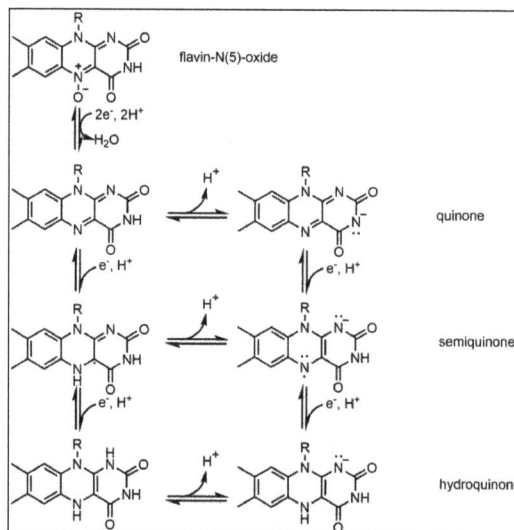

Along with what is seen above, other reactive forms of FAD can be formed and consumed. These reactions involve the transfer of electrons and the making/breaking of chemical bonds. Through reaction mechanisms, FAD is able to contribute to chemical activities within biological systems. The following pictures depict general forms of some of the actions that FAD can be involved in.

Mechanisms 1 and 2 represent hydride gain, in which the molecule gains what amounts to be one hydride ion. Mechanisms 3 and 4 radical formation and hydride loss. Radical species contain unpaired electron atoms and are very chemically active. Hydride loss is the inverse process of the hydride gain seen before. The final two mechanisms show nucleophilic addition and a reaction using a carbon radical.

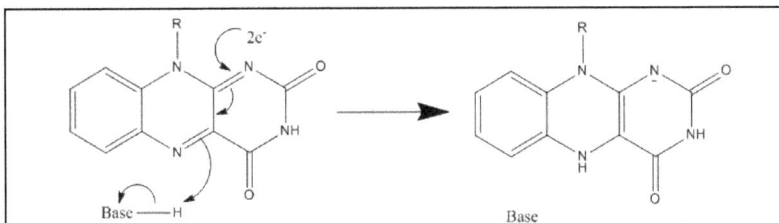

Mechanism 1. Hydride transfer occurs by addition of H⁺ and 2 e⁻

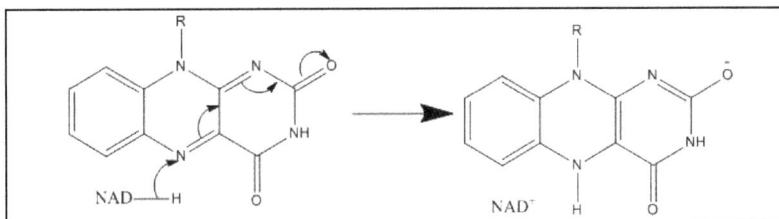

Mechanism 2. Hydride transfer by abstraction of hydride from NADH

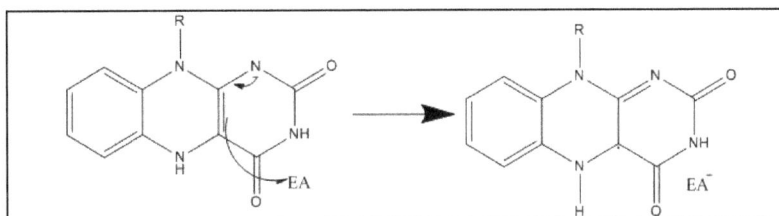

Mechanism 3. Radical formation by electron abstraction

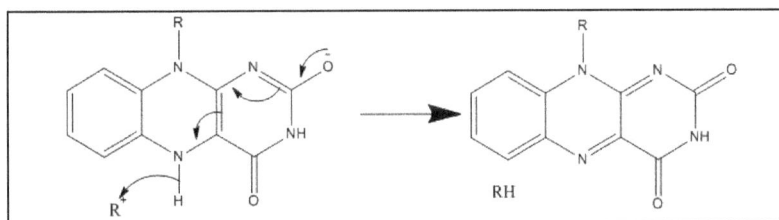

Mechanism 4. The loss of hydride to electron deficient R group

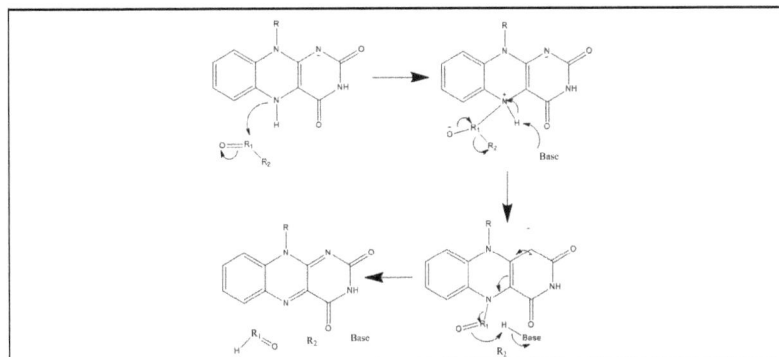

Mechanism 5. Use of nucleophilic addition to break $R_1$-$R_2$ bond

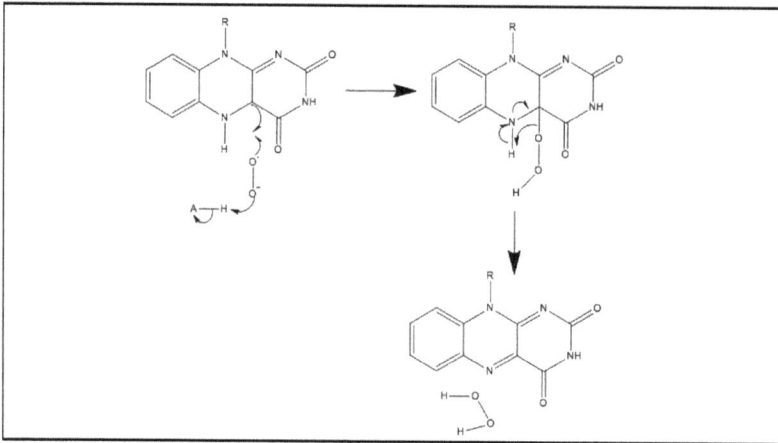

Mechanism 6. Carbon radical reacts with $O_2$ and acid to form $H_2O_2$

## Biosynthesis

FAD plays a major role as an enzyme cofactor along with flavin mononucleotide, another molecule originating from riboflavin.Bacteria, fungi and plants can produce riboflavin, but other eukaryotes, such as humans, have lost the ability to make it. Therefore, humans must obtain riboflavin, also known as vitamin B2, from dietary sources. Riboflavin is generally ingested in the small intestine and then transported to cells via carrier proteins.Riboflavin kinase (EC 2.7.1.26) adds a phosphate group to riboflavin to produce flavin mononucleotide, and then FAD synthetase attaches an adenine nucleotide; both steps require ATP.Bacteria generally have one bi-functional enzyme, but archaea and eukaryotes usually employ two distinct enzymes.Current research indicates that distinct isoforms exist in the cytosol and mitochondria.It seems that FAD is synthesized in both locations and potentially transported where needed.

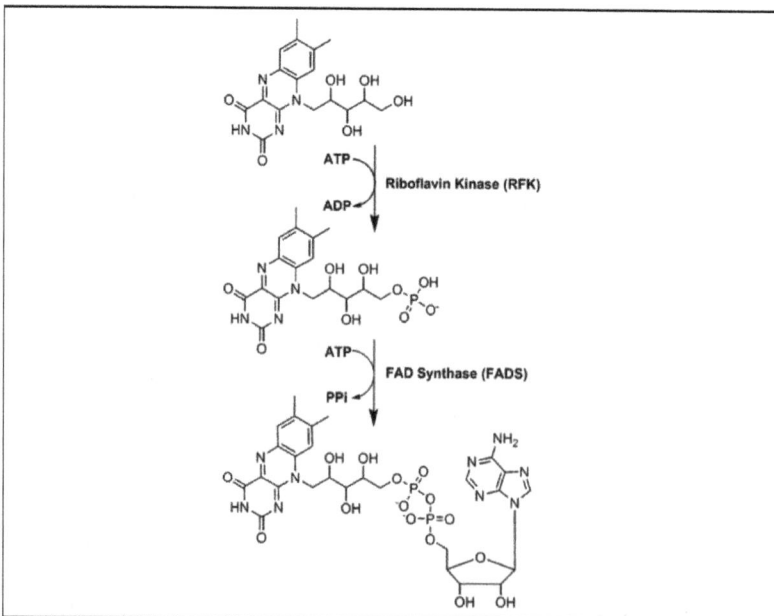

# Function

Flavoproteins utilize the unique and versatile structure of flavin moieties to catalyze difficult redox reactions. Since flavins have multiple redox states they can participate in processes that involve the transfer of either one or two electrons, hydrogen atoms, or hydronium ions. The N5 and C4a of the fully oxidized flavin ring are also susceptible to nucleophilic attack. This wide variety of ionization and modification of the flavin moiety can be attributed to the isoalloxazine ring system and the ability of flavoproteins to drastically perturb the kinetic parameters of flavins upon binding, including flavin adenine dinucleotide (FAD).

The number of flavin-dependent protein encoded genes in the genome (the flavoproteome) is species dependent and can range from 0.1% - 3.5%, with humans having 90 flavoprotein encoded genes.FAD is the more complex and abundant form of flavin and is reported to bind to 75% of the total flavoproteome and 84% of human encoded flavoproteins.Cellular concentrations of free or non-covalently bound flavins in a variety of cultured mammalian cell lines were reported for FAD (2.2-17.0 amol/cell) and FMN (0.46-3.4 amol/cell).

FAD has a more positive reduction potential than $NAD^+$ and is a very strong oxidizing agent. The cell utilizes this in many energetically difficult oxidation reactions such as dehydrogenation of a C-C bond to an alkene. FAD-dependent proteins function in a large variety of metabolic pathways including electron transport, DNA repair, nucleotide biosynthesis, beta-oxidation of fatty acids, amino acid catabolism, as well as synthesis of other cofactors such as CoA, CoQ and heme groups. One well-known reaction is part of the citric acid cycle (also known as the TCA or Krebs cycle); succinate dehydrogenase (complex II in the electron transport chain) requires covalently bound FAD to catalyze the oxidation of succinate to fumarate by coupling it with the reduction of ubiquinone to ubiquinol. The high-energy electrons from this oxidation are stored momentarily by reducing FAD to $FADH_2$. $FADH_2$ then reverts to FAD, sending its two high-energy electrons through the electron transport chain; the energy in $FADH_2$ is enough to produce 1.5 equivalents of ATP by oxidative phosphorylation. There are also redox flavoproteins that non-covalently bind to FAD like Acetyl-CoA-dehydrogenases which are involved in beta-oxidation of fatty acids and catabolism of amino acids like leucine (isovaleryl-CoA dehydrogenase), isoleucine, (short/branched-chain acyl-CoA dehydrogenase), valine (isobutyryl-CoA dehydrogenase), and lysine (glutaryl-CoA dehydrogenase).Additional examples of FAD-dependent enzymes that regulate metabolism are glycerol-3-phosphate dehydrogenase (triglyceride synthesis) and xanthine oxidase involved in purine nucleotide catabolism.There are other noncatalytic roles that FAD can play in flavoproteins such as structural roles, or involved in blue-sensitive light photoreceptors that regulate biological clocks and development, generation of light in bioluminescent bacteria.

# Flavoproteins

Flavoproteins have either an FMN or FAD molecule as a prosthetic group, this prosthetic group can be tightly bound or covalently linked. Only about 5-10% of flavoproteins have a covalently linked FAD, but these enzymes have stronger redox power. In some instances, FAD can provide structural support for active sites or provide stabilization of intermediates during catalysis. Based on the available structural data, the known FAD-binding sites can be divided into more than 200 different types.

There are 90 flavoproteins in the human genome; about 84% require FAD, and around 16% require FMN, whereas 5 proteins require both to be present. Flavoproteins are mainly located in the mitochondria because of their redox power. Of all flavoproteins, 90% perform redox reactions and the other 10% are transferases, lyases, isomerases, ligases.

## Oxidation of Carbon-heteroatom Bonds

## Carbon-nitrogen

Monoamine oxidase (MAO) is an extensively studied flavoenzyme due to its biological importance with the catabolism of norepinephrine, serotonin and dopamine. MAO oxidizes primary, secondary and tertiary amines, which nonenzymatically hydrolyze from the imine to aldehyde or ketone. Even though this class of enzyme has been extensively studied, its mechanism of action is still being debated. Two mechanisms have been proposed: a radical mechanism and a nucleophilic mechanism. The radical mechanism is less generally accepted because there is currently no spectral or electron paramagnetic resonance evidence to show the presence of a radical intermediate. The nucleophilic mechanism is more favored because it is supported by site-directed mutagenesis studies which mutated two tyrosine residues that were expected to increase the nucleophilicity of the substrates.

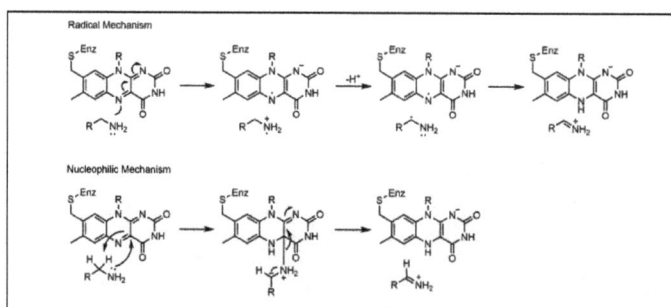

## Carbon-oxygen

Glucose oxidase (GOX) catalyzes the oxidation of β-D-glucose to D-glucono-δ-lactone with the simultaneous reduction of enzyme-bound flavin. GOX exists as a homodimer, with each subunit binding one FAD molecule. Crystal structures show that FAD binds in a deep pocket of the enzyme near the dimer interface. Studies showed that upon replacement of FAD with 8-hydroxy-5-carba-5-deaza FAD, the stereochemistry of the reaction was determined by reacting with the *re* face of the flavin. During turnover, the neutral and anionic semiquinones are observed which indicates a radical mechanism.

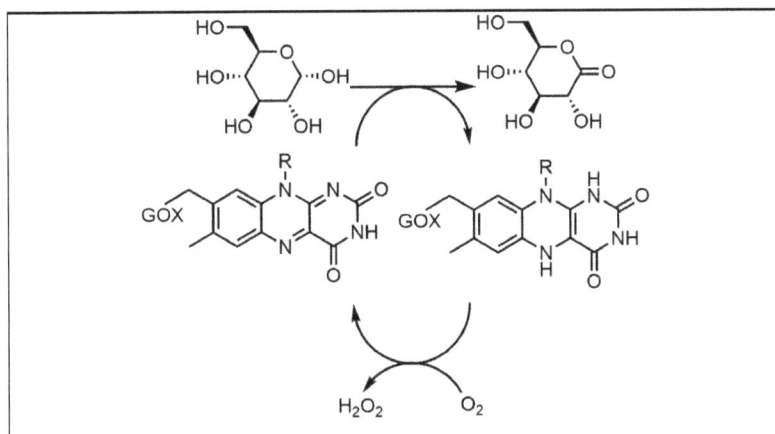

## Carbon-sulfur

Prenylcysteine lyase (PCLase) catalyzes the cleavage of prenylcysteine (a protein modification) to form an isoprenoid aldehyde and the freed cysteine residue on the protein target. The FAD is non-covalently bound to PCLase. Not many mechanistic studies have been done looking at the reactions of the flavin, but the proposed mechanism is shown below. It is proposed that there is a hydride transfer from the C1 of the prenyl moiety to FAD that results in the reduction of the flavin to $FADH_2$ and the formation of a carbocation that is stabilized by the neighboring sulfur atom. $FADH_2$ then reacts with molecular oxygen to restore the oxidized enzyme.

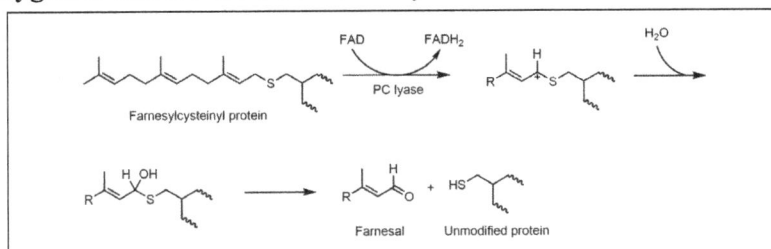

## Carbon-carbon

UDP-N-acetylenolpyruvylglucosamine Reductase (MurB) is an enzyme that catalyzes the NADPH-dependent reduction of enolpyruvyl-UDP-N-acetylglucosamine (substrate) to the corresponding D-lactyl compound UDP-N-acetylmuramic acid (product). MurB is a monomer and contains one FAD molecule. Before the substrate can be converted to product, NADPH must first reduce FAD. Once $NADP^+$ dissociates, the substrate can bind and the reduced flavin can reduce the product.

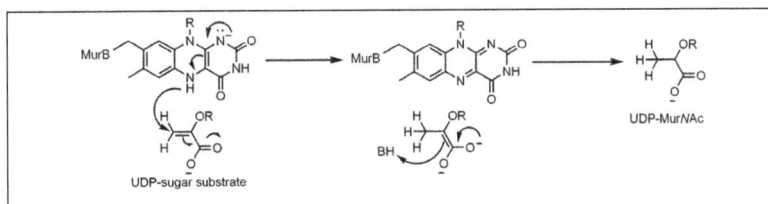

## Thiol/Disulfide Chemistry

Glutathione reductase (GR) catalyzes the reduction of glutathione disulfide (GSSG) to glutathione (GSH). GR requires FAD and NADPH to facilitate this reaction; first a hydride must be transferred from NADPH to FAD. The reduced flavin can then act as a nucleophile to attack the disulfide, this forms the C4a-cysteine adduct. Elimination of this adduct results in a flavin-thiolate charge-transfer complex.

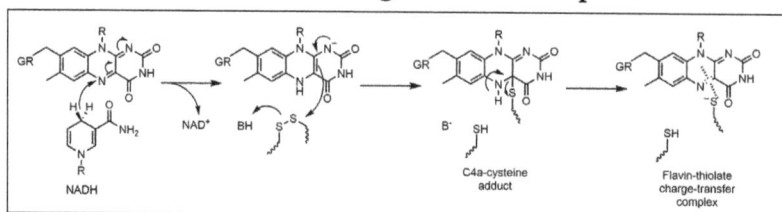

## Electron Transfer Reactions

Cytochrome P450 type enzymes that catalyze monooxygenase (hydroxylation) reactions are dependent on the transfer of two electrons from FAD to the P450. In eukaryotic cells there are two types of P450 systems. The P450 systems that are located in the endoplasmic reticulum are dependent on a cytochrome P-450 reductase (CPR) that contains both an FAD and an FMN. The two electrons on reduced FAD ($FADH_2$) are transferred one at a time to FMN and then a single electron is passed from FMN to the heme of the P450.

The P450 systems that are located in the mitochondria are dependent on two electron transfer proteins: An FAD containing adrenodoxin reductase (AR) and a small iron-sulfur group containing protein named adrenodoxin. FAD is embedded in the FAD-binding domain of AR. The FAD of AR is reduced to $FADH_2$ by transfer of two electrons from NADPH that binds in the NADP-binding domain of AR. The structure of this enzyme is highly conserved to maintain precisely the alignment of electron donor NADPH and acceptor FAD for efficient electron transfer. The two electrons in reduced FAD are transferred one a time to adrenodoxin which in turn donates the single electron to the heme group of the mitochondrial P450.

The structures of the reductase of the microsomal versus reductase of the mitochondrial P450 systems are completely different and show no homology.

## Redox

p-Hydroxybenzoate hydroxylase (PHBH) catalyzes the oxygenation of p-hydroxybenzoate (pOHB) to 3,4-dihyroxybenzoate (3,4-diOHB); FAD, NADPH and molecular oxygen are all required for this reaction. NADPH first transfers a hydride equivalent to FAD, creating $FADH^-$, and then $NADP^+$ dissociates from the enzyme. Reduced PHBH then reacts with molecular oxygen to form the flavin-C(4a)-hydroperoxide. The flavin hydroperoxide quickly hydroxylates pOHB, and then eliminates water to regenerate oxidized flavin. An alternative flavin-mediated oxygenation mechanism involves the use of a flavin-N(5)-oxide rather than a flavin-C(4a)-(hydro)peroxide.

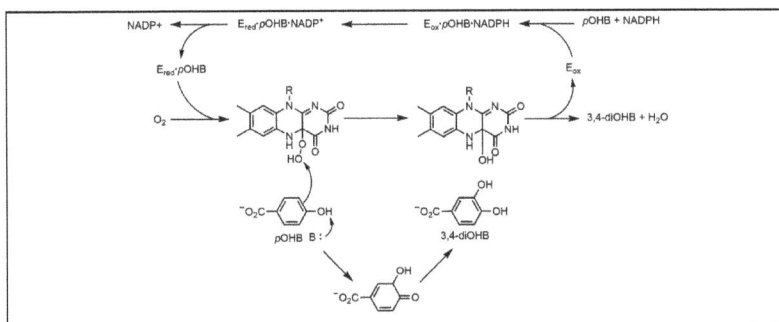

## Nonredox

Chorismate synthase (CS) catalyzes the last step in the shikimate pathway—the formation of chorismate. There are two classes of CS, both of which require FMN, but are divided on their need for NADPH as a reducing agent. The proposed mechanism for CS involves radical species. The radical flavin species has not been detected spectroscopically without using a substrate analogue, which suggests that it is short-lived. However, when using a fluorinated substrate, a neutral flavin semiquinone was detected.

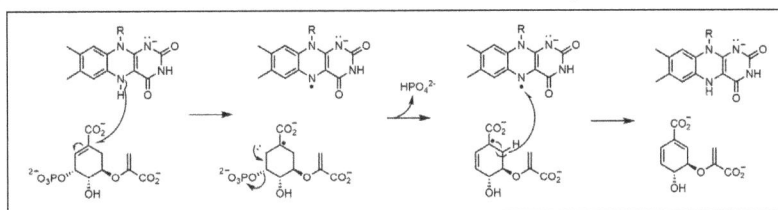

## Complex Flavoenzymes

Glutamate synthase catalyzes the conversion of 2-oxoglutarate into L-glutamate with L-glutamine serving as the nitrogen source for the reaction. All glutamate synthases are iron-sulfur flavoproteins containing an iron-sulfur cluster and FMN. The three classes of glutamate synthases are categorized based on their sequences and biochemical properties. Even though there are three classes of this enzyme, it is believed that they all operate through the same mechanism, only differing by what first reduces the FMN. The enzyme produces two glutamate molecules: one by the hydrolysis of glutamine (forming glutamate and ammonia), and the second by the ammonia produced from the first reaction attacking 2-oxoglutarate, which is reduced by FMN to glutamate.

## Clinical Significance

### Flavoprotein-related Diseases

Due to the importance of flavoproteins, it is unsurprising that approximately 60% of human flavoproteins cause human disease when mutated. In some cases, this is due to a decreased affinity for FAD or FMN and so excess riboflavin intake may lessen disease symptoms, such as for multiple acyl-CoA dehydrogenase deficiency. In addition, riboflavin deficiency itself (and the resulting lack of FAD and FMN) can cause health issues.For example, in ALS patients, there are decreased levels of FAD synthesis. Both of these paths can result in a variety of symptoms, including developmental or gastrointestinal abnormalities, faulty fat break-down, anemia, neurological problems, cancer or heart disease, migraine, worsened vision and skin lesions. The pharmaceutical industry therefore produces riboflavin to supplement diet in certain cases. In 2008, the global need for riboflavin was 6,000 tons per year, with production capacity of 10,000 tons. This $150 to 500 million market is not only for medical applications, but is also used as a supplement to animal food in the agricultural industry and as a food colorant.

## Drug Design

New design of anti-bacterial medications is of continuing importance in scientific research as bacterial antibiotic resistance to common antibiotics increases. A specific metabolic protein that uses FAD (Complex II) is vital for bacterial virulence, and so targeting FAD synthesis or creating FAD analogs could be a useful area of investigation. Already, scientists have determined the two structures FAD usually assumes once bound: either an extended or a butterfly conformation, in which the molecule essentially folds in half, resulting in the stacking of the adenine and isoalloxazine rings. FAD imitators that are able to bind in a similar manner but do not permit protein function could be useful mechanisms of inhibiting bacterial infection. Alternatively, drugs blocking FAD synthesis could achieve the same goal; this is especially intriguing because human and bacterial FAD synthesis relies on very different enzymes, meaning that a drug made to target bacterial FAD synthase would be unlikely to interfere with the human FAD synthase enzymes.

## Optogenetics

Optogenetics allows control of biological events in a non-invasive manner. The field has advanced in recent years with a number of new tools, including those to trigger light sensitivity, such as the Blue-Light-Utilizing FAD domains (BLUF). BLUFs encode a 100 to 140 amino acid sequence that was derived from photoreceptors in plants and bacteria. Similar to other photoreceptors, the light causes structural changes in the BLUF domain that results in disruption of downstream interactions. Current research investigates proteins with the appended BLUF domain and how different external factors can impact the proteins.

## Treatment Monitoring

There are a number of molecules in the body that have native fluorescence including tryptophan, collagen, FAD, NADH and porphyrins. Scientists have taken advantage of this by using them to monitor disease progression or treatment effectiveness or aid in diagnosis. For instance, native fluorescence of a FAD and NADH is varied in normal tissue and oral submucous fibrosis, which is an early sign of invasive oral cancer. Doctors therefore have been employing fluorescence to assist in diagnosis and monitor treatment as opposed to the standard biopsy.

## Nicotinamide Adenine Dinucleotide

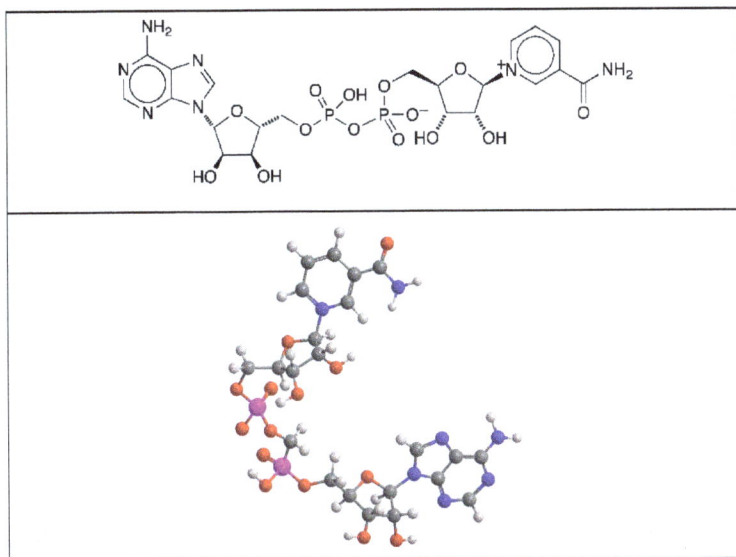

Nicotinamide adenine dinucleotide (abbreviated NAD and sometimes called nadide) is a biomolecule present in all living cells. As its name implies, it consists of two nucleotides, one with an adenine base and the other with a nicotinamide base. The two are joined by their phosphate groups.

NAD exists in an oxidized form (NAD$^+$) and a reduced form (NADH). The neutral form of NAD$^+$ is shown in the images. This is the more common β-isomer; the α-isomer has the opposite stereochemistry at the nicotinamide linkage.

The most important function of NAD$^+$ is as an oxidizing agent in cellular metabolism. (The reduced form NADH, conversely, acts as a metabolic reducing agent.) NAD$^+$ also has roles in adenosine diphosphate (ADP)–ribose transfer reactions that involve the poly(ADP–ribose) polymerase enzyme1 (PARP1) and several other enzymatic processes.

## Role of NAD$^+$

One role of NAD$^+$ is to initiate the electron transport chain by the reaction with an organic metabolite (intermediate in metabolic reactions). This is an oxidation reaction where 2 hydrogen atoms (or 2 hydrogen ions and 2 electrons) are removed from the organic metabolite. (The organic metabolites are usually from the citric acid cycle and the oxidation of fatty acids--details in following pages.) The reaction can be represented simply where M = any metabolite.

$$MH_2 + NAD^+ \rightarrow NADH + H^+ + M :+ \text{energy}$$

One hydrogen is removed with 2 electrons as a hydride ion ( $H^-$ ) while the other is removed as the positive ion ( $H^+$ ). Usually the metabolite is some type of alcohol which is oxidized to a ketone.

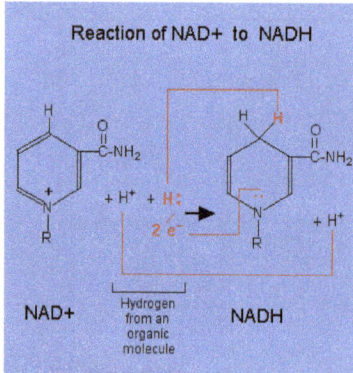

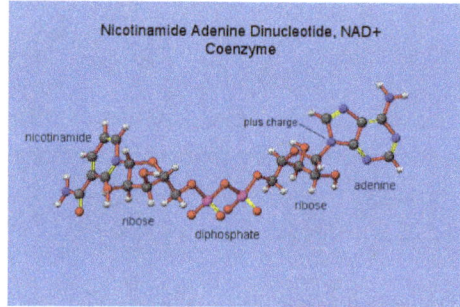

## Alcohol Dehydrogenase

The NAD$^+$ is represented as cyan in figure The alcohol is represented by the space filling red, gray, and white atoms. The reaction is to convert the alcohol, ethanol, into ethanal, an aldehyde.

$$CH_3CH_2OH + NAD+ \rightarrow CH_3CH = O + NADH + H^+$$

This is an oxidation reaction and results in the removal of two hydrogen ions and two electrons which are added to the NAD$^+$, converting it to NADH and H$^+$. This is the first reaction in the metabolism of alcohol. The active site of ADH has two binding regions. The coenzyme binding site, where NAD$^+$ binds, and the substrate binding site, where the alcohol binds. Most of the binding site for the NAD$^+$ is hydrophobic as represented in green. Three key amino acids involved in the catalytic oxidation of alcohols to aldehydes and ketones. They are ser-48, phe 140, and phe 93.

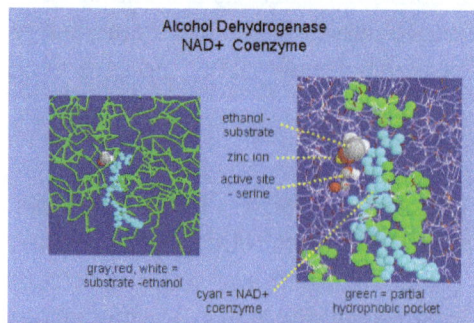

Active site of Alcohol Dehydrogenase

# Heme

A heme is an organic, ring-shaped molecule. Due to its special structure, a heme is capable of holding, or "hosting" an iron molecule. A heme is made from 4 pyrroles, which are small pentagon-shaped molecules made from 4 carbons and 1 nitrogen. Four pyrroles together form a tetrapyrrole. If the tetrapyrrole has substitutions on the side chains which allow it to hold a metal ion, it is called a porphyrin. Thus, a heme is an iron-holding porphyrin.

The iron molecule in a heme is held in place by the balanced attractive forces of the four nitrogen molecules. The nitrogen molecules all point toward the inside of the larger ring they create. The double and single bonds which connect the pyrroles are arranged evenly, so that the electrons stay balanced and the entire molecule remains stable. This makes it an aromatic molecule. A heme molecule can be seen below.

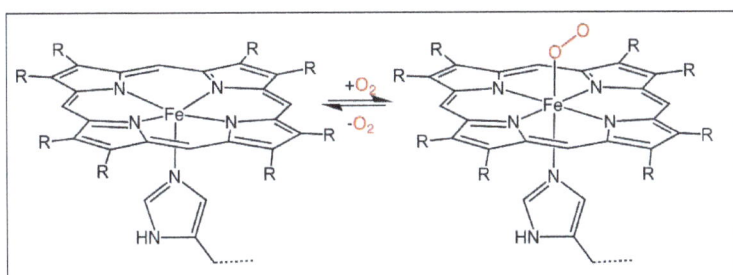

Hemes are used for two known reasons: to carry oxygen and to transport or store electrons. In the above image, you can see how gaseous oxygen can reversibly bind to the heme complex. Organisms use the heme molecule, in complex with specially-shaped proteins, to transport oxygen and move electrons. These special proteins, like hemoglobin and myoglobin, are made to help the heme complex hold or release oxygen at the appropriate times.

The red color of blood is produced by the heme and iron ion interacting to absorb other colors and only reflect red. A somewhat different effect is seen in chlorophyll. Chlorophyll is a porphyrin complex used in photosynthesis. Instead of iron, chlorophyll houses a magnesium ion, and chlorophyll has different side chains than a heme group. This produces the green color of plants, rather than the reds and purples of blood.

## Heme Structure

Like all porphyrins, heme has a base structure of a large ring of four pyrroles. This base molecule, only seen rarely as an intermediate in nature, is called porphin. There are many different forms of heme, which correspond to the many functions it must serve in an organism. Specific proteins use the varying side chains to attach to, and they change the properties of the heme. However, the base structure is always the same. It is the tetrapyrrole shown below.

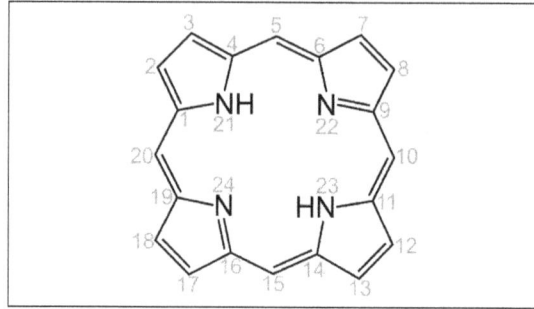

The numbers on the molecules indicate points in which the molecule may receive substitutions and be modified for a specific use. Differences in the side-chains attached to carbons 3, 8, and 18 constitute the difference between some of the most common heme groups. For instance, hemoglobin and myoglobin both carry Heme B. Heme B carries oxygen, and the proteins it is attached to help it release the oxygen at the appropriate time. Heme A, on the other hand, works in the electron transport chain as part of cytochrome c. This means that it is involved in transporting proteins and catalyzing reactions. The only difference between the two molecules are the side-chains attached at carbons 3 and 18. The side chain on carbon 8 remains the same.

## Heme Function

Heme has two understood functions. It can bind gases, such as oxygen, and transport them throughout an organism. Special proteins then force the heme to release its oxygen at the appropriate time. A good example of a protein of this type is hemoglobin. Hemoglobin is found in all blood cells, attached to the cell membrane, exposing the heme group to the blood plasma. Thus, when the blood cells pass through the lungs, they bind up as much oxygen as the iron in the heme can handle.

The blood cells then travel to various parts of the body, such as the muscles. These cells are actively using up oxygen and releasing carbon dioxide as a byproduct. Carbon dioxide forms an acid in the blood plasma, lowering the pH of the blood. Like all proteins, hemoglobin reacts to changes in pH by changing shape. This change in shape forces the oxygen off of the heme complex, releasing the oxygen into the blood plasma. The oxygen diffuses into the muscle cells, where it is bound by myoglobin and transported to the mitochondria to be used. Myoglobin also has a heme group, but it operates in a different way so that oxygen remains bound until reaching the mitochondria.

The second function of hemes, holding electrons and facilitating reactions in the electron transport chain, occurs in all organisms. During oxidative phosphorylation in the mitochondrial membrane, electrons must be passed down a series of reactions, which slowly extract their energy before depositing them in water and carbon dioxide. The energy gained is stored in the bonds of the molecule ATP, which most organisms use as a primary source of energy. The heme groups in these cytochromes are different than those in hemoglobin, for they have different functions and bind to different proteins.

# Folate

Folate is a water-soluble B-vitamin, which is also known as vitamin B9 or folacin. Naturally occurring folates exist in many chemical forms; folates are found in food, as well as in metabolically active forms in the human body. Folic acid is the major synthetic form found in fortified foods and vitamin supplements. Other synthetic forms include folinic acid (Figure) and levomefolic acid. Folic acid has no biological activity unless converted into folates (1). In the following discussion, forms found in food or the body are referred to as "folates," while the form found in supplements or fortified food is referred to as "folic acid."

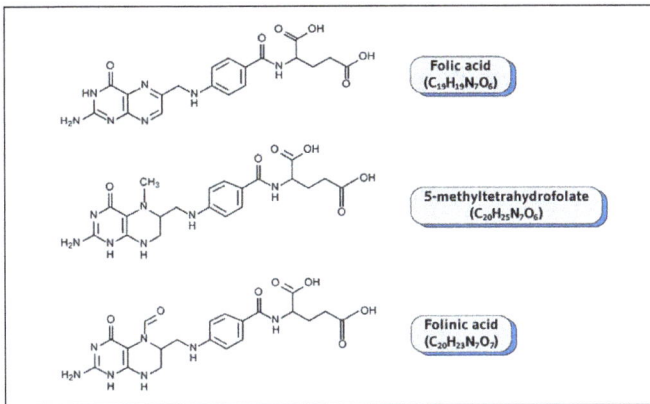

Figure: Chemical Structures

# Function

## One-carbon Metabolism

The only function of folate coenzymes in the body appears to be in mediating the transfer of one-carbon units (2). Folate coenzymes act as acceptors and donors of one-carbon units in a variety of reactions critical to the metabolism of nucleic acids and amino acids (Figure).

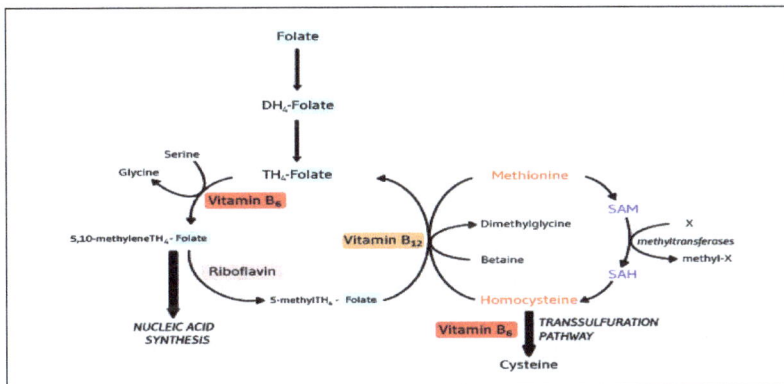

Figure: Overview of One- carbon Metabolism

5, 10-methylenetetrahydrofolate is required for the synthesis of nucleic acids, and 5- methyltetrahydrofolate is required for the formation of methionine from homocysteine. Methionine, in the form of methyl donor S-adenosylmethionine (SAM), is essential to many biological methylation reactions, including DNA methylation. Methylenetetrahydrofolate reductase (MTHER) is a riboflavin (FAD)-dependent enzyme that catalyzes the reduction of 5,10-methylenetetrahydrofolate to 5-methyltetrahydrofolate. SAM, S-adenosylmethionine; SAH, S-adenosylhomocysteine; $TH_4$-Folate, Tetrahydrofolate.

## Nucleic Acid Metabolism

Folate coenzymes play a vital role in DNA metabolism through two different pathways. (1) The synthesis of DNA from its precursors (thymidine and purines) is dependent on folate coenzymes. (2) A folate coenzyme is required for the synthesis of methionine from homocysteine, and methionine is required for the synthesis of S-adenosylmethionine (SAM). SAM is a methyl group (one-carbon unit) donor used in most biological methylation reactions, including the methylation of a number of sites within DNA, RNA, proteins, and phospholipids. The methylation of DNA plays a role in controlling gene expression and is critical during cell differentiation. Aberrations in DNA methylation have been linked to the development of cancer.

## Amino Acid Metabolism

Folate coenzymes are required for the metabolism of several important amino acids, namely methionine, cysteine, serine, glycine, and histidine. The synthesis of methionine from homocysteine is catalyzed by methionine synthase, an enzyme that requires not only folate (as 5-methyltetrahydrofolate) but also vitamin B12. Thus, folate (and/or vitamin B12) deficiency can result in decreased synthesis of methionine and an accumulation of homocysteine. Elevated blood concentrations of homocysteine have been considered for many years to be a risk factor for some chronic diseases, including cardiovascular disease and dementia.

## Nutrient interactions

## Vitamin $B_{12}$ and vitamin $B_6$

The metabolism of homocysteine, an intermediate in the metabolism of sulfur-containing amino acids, provides an example of the interrelationships among nutrients necessary for optimal physiological function and health. Healthy individuals utilize two different pathways to metabolize homocysteine (Figure). One pathway (methionine synthase) synthesizes methionine from homocysteine and is dependent on both folate and vitamin B12 as cofactors. The other pathway converts homocysteine to another amino acid, cysteine, and requires two vitamin B6-dependent enzymes. Thus, the concentration of homocysteine in the blood is regulated by three B-vitamins: folate, vitamin B12, and vitamin B6 (4). In some individuals, riboflavin (vitamin B2) is also involved in the regulation of homocysteine concentrations.

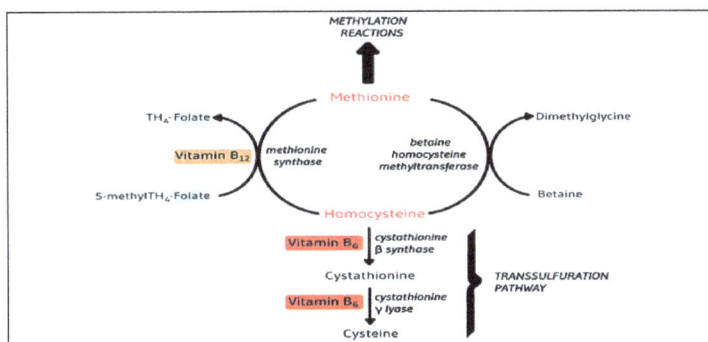

Homocysteine Metabolism

Homocysteine is methylated to form the essential amino acid methionine in two pathways. The reaction of homocysteine remethylation catalyzed by the vitamin $B_{12}$-dependent methionine sysnthase captures a methyl group from the folate-dependent one-carbon pool (5-methyltetrahydrofolate). A second pathway requires betaine (N,N,N-trimethylglycine) as a methyltransferase. The catabolic pathway of homocysteine, known as the transsulfuration pathway, converts homocysteine to the amino acid cysteine via two vitamin $B_6$ (PLP)-dependent enzymes. Cystathionine $\beta$ synthase catalyzes the condensation of homocysteine with serine to form cystathionine, and cystathionine is then converted to cysteine, $\alpha$ -ketobutyrate, and ammonia by cystathionine $\lambda$ lyase. $TH_4$- Folate, Tetrahydrofolate.

## Riboflavin

Although less well recognized, folate has an important metabolic interaction with riboflavin. Riboflavin is a precursor of flavin adenine dinucleotide (FAD), a coenzyme required for the activity of the folate-metabolizing enzyme, 5,10-methylenetetrahydrofolate reductase (MTHFR). FAD-dependent MTHFR in turn catalyzes the reaction that generates 5-methyltetrahydrofolate. This active form of folate is required to form methionine from homocysteine. Along with other B-vitamins, higher riboflavin intakes have been associated with decreased plasma homocysteine concentrations. The effects of riboflavin on folate metabolism appear to be greatest in individuals homozygous for the common c.677C>T polymorphism (i.e., TT genotype) in the MTHFR gene. These individuals (about 10% of adults worldwide) typically present with low folate status, along with elevated homocysteine concentrations, particularly when folate and/or riboflavin intake is suboptimal. The elevated homocysteine concentration in these individuals, however, is highly responsive to lowering with riboflavin supplementation, confirming the importance of the riboflavin-MTHFR interaction.

## Vitamin C

Vitamin C may limit degradation of natural folate coenzymes and supplemental folic acid in the stomach and thus improve folate bioavailability. A cross-over trial in nine healthy men found that oral co-administration of 5-methyltetrahydrofolic acid (343 µg) and vitamin C (289 mg or 974 mg) was associated with higher concentrations of serum folate compared to 5-methyltetrahydrofolic acid alone (8). Moreover, a recent study suggested that several genetic variations of folate metabolism might influence the effect of vitamin C on folate metabolism (9).

## Bioavailability

Dietary folates exist predominantly in the polyglutamyl form (containing several glutamate residues), whereas folic acid—the synthetic vitamin form—is a monoglutamate, containing just one glutamate moiety. In addition, natural folates are reduced molecules, whereas folic acid is fully oxidized. These chemical differences have major implications for the bioavailability of the vitamin such that folic acid is considerably more bioavailable than naturally occurring food folates at equivalent intake levels.

The intestinal absorption of dietary folates is a two-step process that involves the hydrolysis of folate polyglutamates to the corresponding monoglutamyl derivatives, followed by their transport into intestinal cells. There, folic acid is converted into a naturally occurring folate, namely 5-methyltetrahydrofolate, which is the major circulating form of folate in the human body.

The bioavailability of naturally occurring folates is inherently limited and variable. There is much variability in the ease with which folates are released from different food matrices, and the polyglutamyl "tail" is removed (de-conjugation) before uptake by intestinal cells. Also, other dietary constituents can contribute to instability of labile folates during the processes of digestion. As a result, naturally occurring folates show incomplete bioavailability compared with folic acid. The bioavailability of folic acid, in contrast, is assumed to be 100% when ingested as a supplement, while folic acid in fortified food is estimated to have about 85% the bioavailability of supplemental folic acid.

## Transport

Folate and its coenzymes require transporters to cross cell membranes. Folate transporters include the reduced folate carrier (RFC), the proton-coupled folate transporter (PCFT), and the folate receptor proteins, FRα and FRβ. Folate homeostasis is supported by the ubiquitous distribution of folate transporters, although abundance and importance vary among tissues (10). PCFT plays a major role in folate intestinal transport since mutations affecting the gene encoding PCFT cause hereditary folate malabsorption. Defective PCFT also leads to impaired folate transport into the brain. FRα and RFC are also critical for folate transport across the blood-brain barrier when extracellular folate is either low or high, respectively. Folate is essential for the proper development of the embryo and the fetus. The placenta is known to concentrate folate to the fetal circulation, leading to higher folate concentrations in the fetus compared to those found in the pregnant woman. All three types of receptors have been associated with folate transport across the placenta during pregnancy.

## Biotin

Biotin, also known as vitamin $B_7$ or vitamin H, is one of the B vitamins, a group of chemically distinct, water-soluble vitamins that also includes thiamine, riboflavin, niacin, pantothenic acid, pyridoxine, folic acid, and others. Vitamins are organic (carbon-containing) nutrients obtained through the diet and essential in small amounts for normal metabolic reactions in humans. The B vitamins (vitamin B complex) were once considered to be a single vitamin, like vitamin C. However, vitamin B is now seen as a complex of different vitamins that generally are found in the same foods.

Biotin is important in a number of essential metabolic reactions in humans, including catalyzing the synthesis of fatty acids, metabolism of the amino acid leucine, and gluconeogenesis (generation of glucose from non-sugar carbon substrates like pyruvate, glycerol, and amino acids). Biotin is important in cell growth; plays a role in the Krebs cycle, which is the biochemical pathway in which energy is released from food (glucose, amino acids, and fat); helps with the transfer of carbon dioxide; and is useful in maintaining a steady blood sugar level.

A harmonious relationship with symbiotic bacteria in the intestine of humans helps in preventing biotin deficiency as these bacteria synthesize small amounts of biotin. On the other hand, biotin reveals the importance of balance in one's diet, as excessive consumption of raw egg-whites over a long period of time can result in biotin deficiency, as a protein in the egg-whites binds with biotin and results in its removal.

## Structure

Biotin has the chemical formula $C_{10}H_{16}N_2O_3S$. Biotin is composed of an ureido (tetrahydroimidizalone) ring fused with a tetrahydrothiophene ring, which is an organic compound consisting of a five-membered ring containing four carbon atoms and a sulfur atom. A valeric acid substituent—straight chain alkyl carboxylic acid with the chemical formula $CH_3(CH_2)_3COOH$)—is attached to one of the carbon atoms of the tetrahydrothiophene ring.

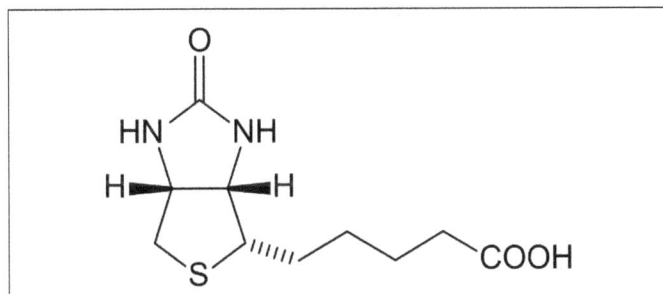

Structure of biotin

## Biotin Deficiency

Biotin deficiency is a rare metabolic genetic disorder. For that reason, statutory agencies in many countries (e.g., the Australian Department of Health and Aging) do not prescribe a recommended daily intake. Biotin deficiency can have a very serious, even fatal, outcome if it is allowed to progress without treatment. Signs and symptoms of biotin deficiency can develop in persons of any age, race, or gender.

Biotin deficiency rarely occurs in healthy individuals, since the daily requirements of biotin are low, many foods contain adequate amounts, intestinal bacteria synthesize small amounts, and the body effectively scavenges and recycles biotin from bodily waste. However, deficiency can be caused by excessive consumption of raw egg-whites over a long period (months to years). Egg-whites contain high levels of avidin, a protein that binds biotin strongly. Once a biotin-avidin complex forms, the bond is essentially irreversible. The biotin-avidin complex is not broken down nor liberated during digestion, and the biotin-avidin complex is lost in the feces. Once cooked, the egg-white avidin becomes denatured and entirely non-toxic.

Initial symptoms of biotin deficiency include:

1.  Dry skin.

2.  Seborrheic dermatitis.

3.  Fungal infections.

4.  Rashes including erythematous periorofacial macular rash.

5.  Fine and brittle hair.

6.  Hair loss or total alopecia.

If left untreated, neurological symptoms can develop, including:

1.  Mild depression, which may progress to profound lassitude and, eventually, to somnolence.

2.  Changes in mental status.

3.  Generalized muscular pains (myalgias).

4.  Hyperesthesias and paresthesias.

The treatment for biotin deficiency is to simply start taking some biotin supplements.

## Uses

Biotin supplements are often recommended as a natural product to counteract the problem of hair loss in both children and adults. There are, however, no studies that show any benefit in any case where the subject is not actually biotin deficient. The signs and symptoms of biotin deficiency include hair loss that progresses in severity to include loss of eye lashes and eye brows in severely deficient subjects. Some shampoos are available that contain biotin, but it is doubtful whether they would have any useful effect, as biotin is not absorbed well through the skin.

Biotin is often recommended for strengthening hair and nails. Consequently, it is found in many cosmetic and health products for the hair and skin.

Children with a rare inherited metabolic disorder called phenylketonuria (PKU; in which one is unable to break down the amino acid phenylalanine) often develop skin conditions such as eczema and seborrheic dermatitis in areas of the body other than the scalp. The scaly skin changes that occur in people with PKU may be related to poor ability to use biotin. Increasing dietary biotin has been known to improve seborrheic dermatitis in these cases.

People with type 2 diabetes often have low levels of biotin. Biotin may be involved in the synthesis and release of insulin. Preliminary studies in both animals and people suggest that biotin may help improve blood sugar control in those with diabetes, particularly type 2 diabetes.

## Biochemistry

Biotin is a cofactor responsible for carbon dioxide transfer in several carboxylase enzymes:

- Acetyl-CoA carboxylase alpha.

- Acetyl-CoA carboxylase beta.

- Methylcrotonyl-CoA carboxylase.

- Propionyl-CoA carboxylase.

- Pyruvate carboxylase.

The attachment of biotin to various chemical sites, called biotinylation, can be used as an important laboratory technique to study various processes including DNA transcription and replication. Biotin itself is known to biotinylate histones, but is not found naturally on DNA.

Biotin binds very tightly to the tetrameric protein streptavidin, with a dissociation constant Kd in the order of $10^{-15}$ mol/L or $4 \times 10^{-14}$. Holmberg et al. the biotin-streptavidin system is the strongest noncovalent biological interaction known. This is often used in different biotechnological applications. Holmberg et al. showed how to utilize high temperatures to efficiently break the interaction without denaturation of the streptavidin.

In the biology laboratory, biotin is sometimes chemically linked, or tagged, to a molecule or protein for biochemical assays. The specificity of the biotin-streptavidin linkage allow use in molecular, immunological, and cellular assays. Since avidin and streptavidin bind preferentially to biotin, biotin-tagged molecules can be extracted from a sample by mixing them with beads covered with avidin or strepavidin, and washing away anything unbound to the beads.

For example, biotin can be tagged onto a molecule of interest (e.g. protein), and this modified molecule will be mixed with a complex mixture of proteins. Avidin or streptavidin beads are added to the mixture, and the biotinylated molecule will bind to the beads. Any other proteins binding to the biotinylated molecule will also stay with the beads. All other unbound proteins can be washed away, and the scientist can use a variety of methods to determine which proteins have bound to the biotinylated molecule.

Biotinylated antibodies are used to capture avidin or streptavidin both the ELISPOT technique (Enzyme-Linked Immunosorbent SPOT, a method for monitoring immune responses in humans and animals) and the ELISA technique (Enzyme-Linked ImmunoSorbent Assay, a biochemical technique used in immunology to detect the presence of an antibody or an antigen in a sample).

## Pyridoxal Phosphate

Pyridoxal 5'-phosphate (P5P) is the active coenzyme form of Vitamin B6 which can be directly utilized by the body without conversion. There are several compounds that are commonly referred to as B6 including Pyridoxine, Pyridoxal, Pyridoxamine, and their respective 5`-phosphate forms. These are considered Pyridoxine vitamers, and our body is capable of interconverting between them. pyridoxine can be converted to pyridoxal 5`-phosphate, and then back into pyridoxine later. Each of these compounds play unique roles in our body, but only P5P has an active coenzyme role. Pyridoxal 5'-phosphate controls the metabolism of amino acids, aids in the production of neurotransmitters, and is a cofactor for more than 150 enzymes in our body.

Many enzymes involved in the metabolism of amino acids make use of PLP as a coenzyme. The basic function of PLP is to act as an 'electron sink', the meaning of which will become clear as we examine in detail the mechanisms of some PLP-dependant reactions. Most of these reactions will be familiar - they include amino acid racemizations, eliminations, additions, and retro-aldol or retro-Claisen cleavages. What will be new is the use of PLP to stabilize the carbanion intermediates in each of these reactions.

## PLP and the Schiff Base Linkage to Lysine

In the first step of virtually all PLP-dependant reactions, the aldehyde group of the coenzyme forms an imine (Schiff base) linkage with a lysine side chain on the enzyme. (In order to help you to focus on the chemistry happening with the amino acid substrates, the PLP molecule in subsequent figures is colored green - it will be very helpful to refer to color figures).

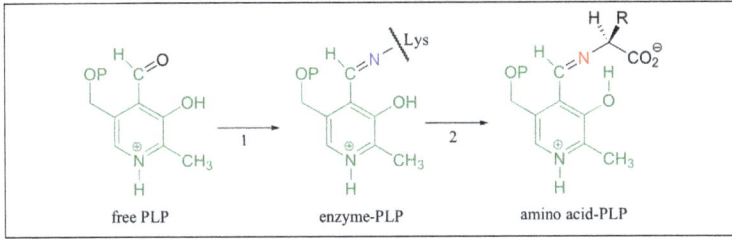

Generally, the next step is an imine exchange, as the amine nitrogen on the amino acid substrate replaces the enzyme lysine nitrogen in the imine linkage. This substrate-co-enzyme adduct is stabilized by a favorable hydrogen bond between the phenol of PLP and the imine nitrogen.

## PLP-dependent Amino Acid Racemases

With these preliminaries accomplished, the real PLP chemistry is ready to start. Let's look first at the reaction catalyzed by PLP-dependent alanine racemase.

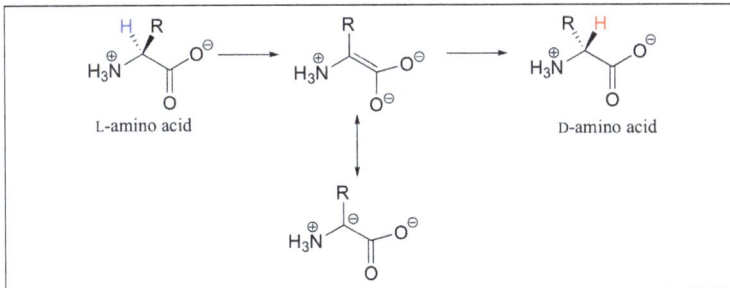

Many other amino acid racemases, however, require the participation of PLP. Once it is linked to PLP in the active site, the a-proton of alanine can be abstracted by an active site base. The negative charge on the carbanion intermediate can be delocalized to the carboxylate group of alanine. The PLP coenzyme, however, provides a much expanded network of conjugated $\pi$-bonds over which the electron density can be delocalized all the way down to the PLP nitrogen.

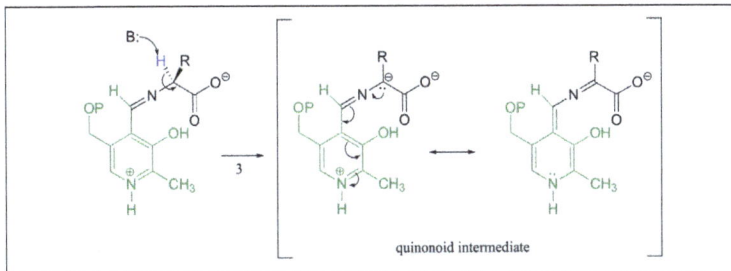

This is what we mean when we say that the job of PLP is to act as an 'electron sink': the coenzyme is extremely efficient at absorbing, or delocalizing, the excess electron density on the deprotonated a-carbon of the reaction intermediate. PLP is helping the enzyme to increase the acidity of the a-hydrogen by stabilizing the conjugate base. A PLP-stabilized carbanion intermediate is commonly referred to as a quinonoid intermediate.

In the remaining steps of the alanine racemase reaction, reprotonation occurs on the opposite side of the substrate, leading to the D-amino acid.

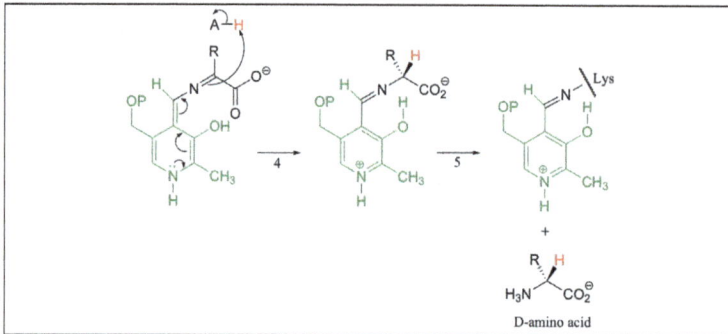

All that remains is another imine exchange, which frees the D-alanine product and re-attaches PLP to the enzymatic lysine side-chain.

## PLP-dependent Decarboxylation

In the alanine racemase reaction above, PLP assisted in breaking the Cα-H bond of the amino acid. Some PLP-dependant enzymes can catalyze the breaking of the bond between Cα and the carboxylate carbon in an amino acid by stabilizing the resulting carbanion intermediate: these are simply decarboxylation reactions. One of the final steps in the synthesis of lysine is a PLP-dependent decarboxylation.

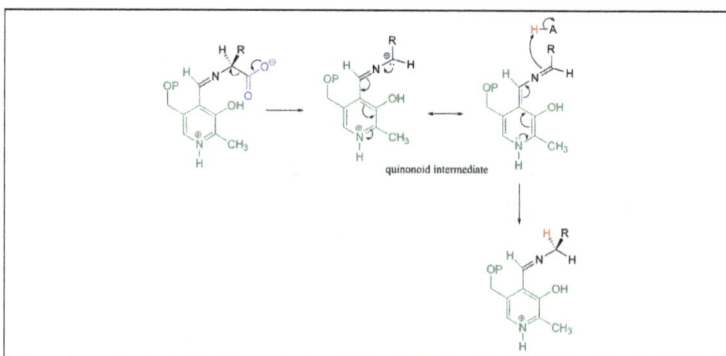

## PLP-dependent Retro-aldol and Retro-claisen Reactions

Still other PLP-dependant enzymes catalyze the breaking of the bond between the alpha-carbon and the first carbon on the amino acid side chain - this is referred to as a Cα-Cβ cleavage, and it actually is a variation on the retro-aldol reaction. When serine is degraded, it is first converted to glycine by serine hydroxymethyltransferase.

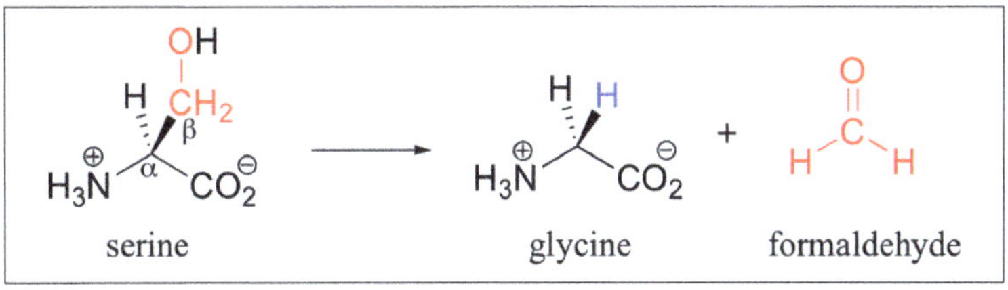

serine                                    glycine              formaldehyde

The retro-aldol step occurs from the PLP-serine Schiff base adduct, as the electrons from the broken bond are stabilized by the electron-sink network of the coenzyme. Just like in a standard retroaldol step, these electrons then go on to form a new carbon-hydrogen bond, leading to a PLP-glycine adduct which is freed by imine exchange.

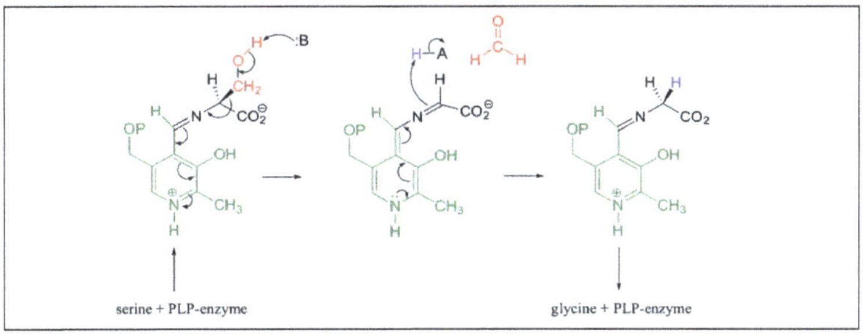

serine + PLP-enzyme                              glycine + PLP-enzyme

PLP also assists in retro-Claisen cleavage reactions, such as this step in the degradation of threonine:

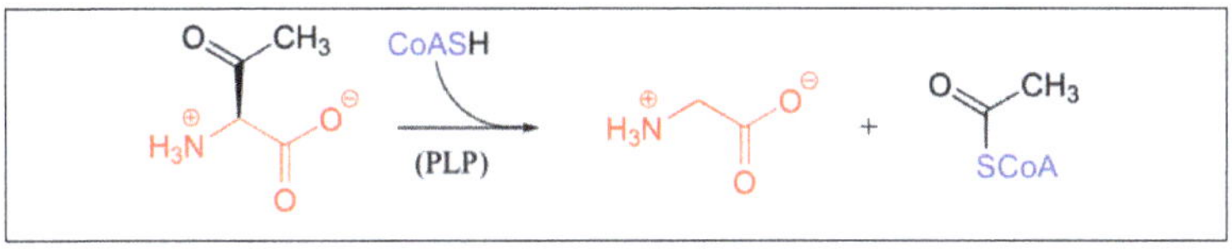

## PLP-dependent Transamination Reactions

One of the most important steps in the degradation of amino acids is elimination of the amino nitrogen atom in the form of urea, which is excreted in the urine. While the metabolic details of this 'urea cycle' are outside of the scope of this text, what is important to understand is that nitrogen atoms from many of the amino acids must be shuttled first to glutamate before being processed for elimination:

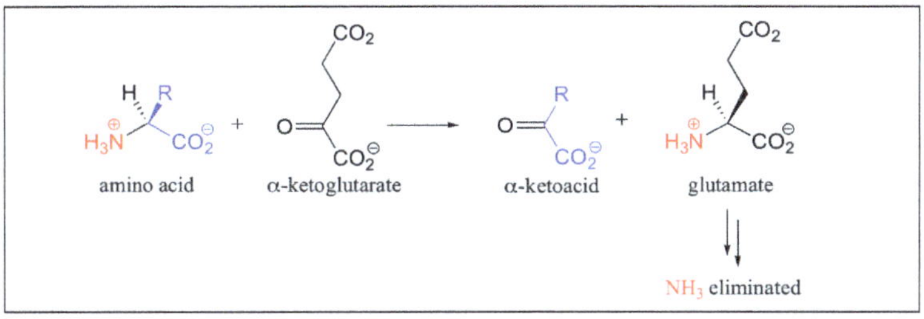

amino acid        α-ketoglutarate        α-ketoacid        glutamate

NH₃ eliminated

The reaction in which the nitrogen group from an amino acid is transferred to alpha-ketoglutarate is accomplished by PLP-dependent enzymes called transaminases. Once again, the first step is abstraction of the alpha-proton from the PLP-amino acid adduct.

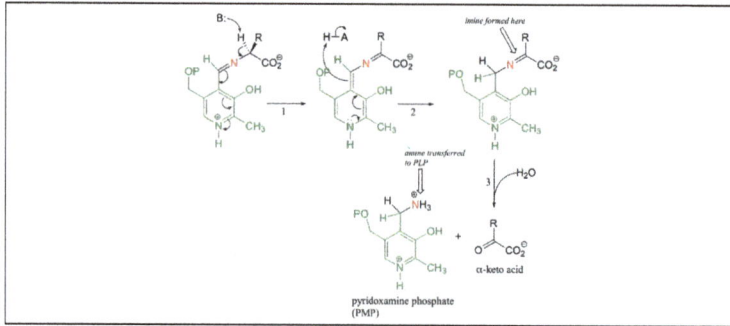

pyridoxamine phosphate
(PMP)

However, in the transaminase reactions this initial deprotonation step is immediately followed by a reprotonation at what was originally the aldehyde carbon of PLP, which results in a new carbon-nitrogen double bond (i.e., an imine) between the a-carbon and the nitrogen of the original amino acid. This imine is then hydrolyzed- this is the step where the nitrogen is removed from the amino acid to form an alpha-keto acid, which can be degraded further.

The coenzyme, which now carries an amine group and is therefore called pyridoxamine phospate (PMP), next transfers the amine group to alpha-ketoglutarate (to form glutamate) through an exact reversal of the whole process we have just seen.

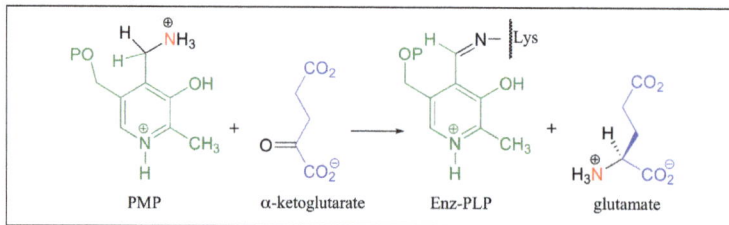

PMP                α-ketoglutarate            Enz-PLP                glutamate

In the overall transaminase reaction, the PLP coenzyme not only provides an electron sink, it also serves as a temporary 'storage space' for a nitrogen atom as it is passed from one amino acid to another.

Transaminase reactions also function in the biosynthesis direction for alanine, aspartate, and glutamate. Alanine, for example, is synthesized from pyruvate by the transfer of an amino group from glutamate.

## PLP-dependent Beta-elimination and Beta-substitution Reactions

Two more reaction types in the PLP toolbox are beta-eliminations and beta-substitutions. Serine dehydratase catalyzes a beta-elimination of the hydroxyl group of serine, leading eventually to pyruvate.

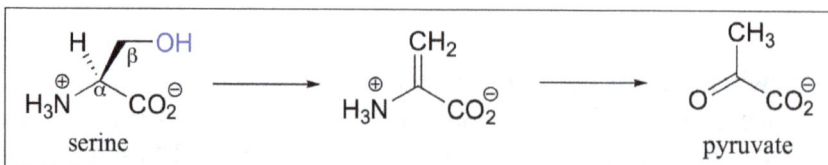

serine                                                                pyruvate

After forming the normal imine linkage with PLP, the a-proton of serine is abstracted by a basic active-site lysine residue, and the coenzyme stabilizes the conjugate base.

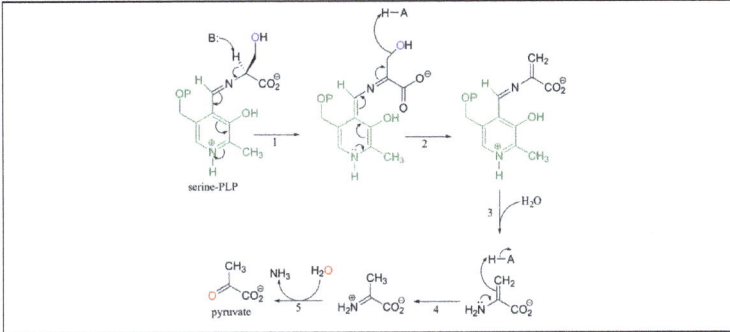

After elimination of water, the dehydrated substrate is freed from the PLP coenzyme by Schiff base transfer, tautomerizes to the imine form, and then is hydrolyzed to pyruvate.

A beta-substitution is simply beta-elimination followed directly by the reverse reaction (Michael addition) with a different nucleophile:

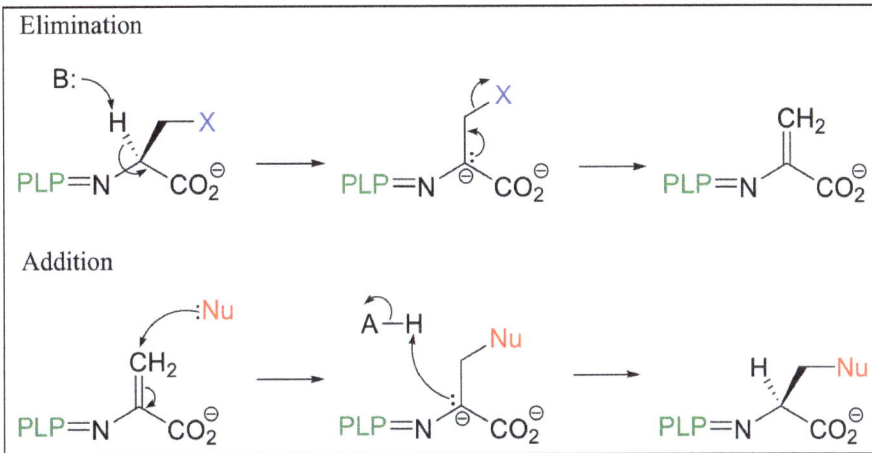

As with virtually all PLP-dependent reactions, the coenzyme serves to stabilize the carbanion intermediates.

In many bacteria, the synthesis of cysteine from serine relies upon a PLP-dependent beta-substitution step. In this pathway, serine is first acetylated by acetyl-CoA (an acyl transfer reaction).

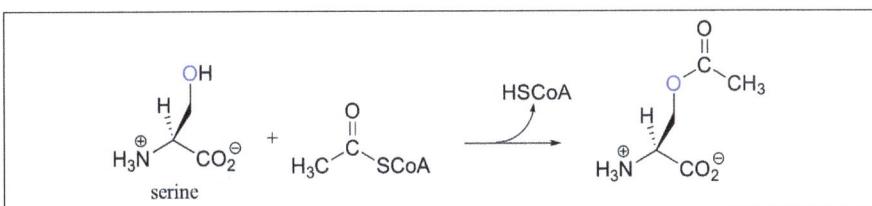

The acetylated serine forms an imine linkage with PLP, then undergoes an elimination in which the acetyl group is expelled (acetyl is, of course, a much weaker base / better leaving group than a hydroxide - thus the function of the initial serine acetylation step).

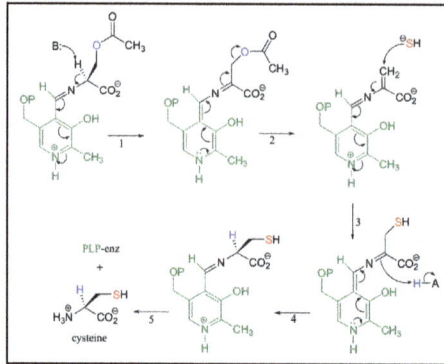

A sulfhydryl ion (SH-) then attacks in a Michael addition, with the intermediate stabilized again by the electron-sink property of PLP. Finally, the cysteine product is released from PLP via an imine exchange reaction with an active site lysine.

## PLP-dependent Gamma-elimination and Gamma-substitution Reactions

The electron sink capability of PLP allows some enzymes to catalyze eliminations at the gamma ($\gamma$) carbon of an amino acid side chain, rather than at the beta-carbon:

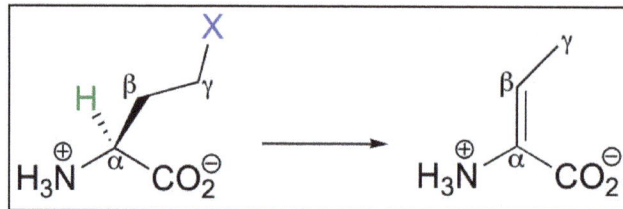

The secret to understanding the mechanism of a gamma-elimination is that PLP acts as an electron sink twice - it absorbs the excess electron density from not one but two proton abstractions. As an example, let's look at the cystathionine gamma-lyase reaction, which is part of the methionine degradation pathway.

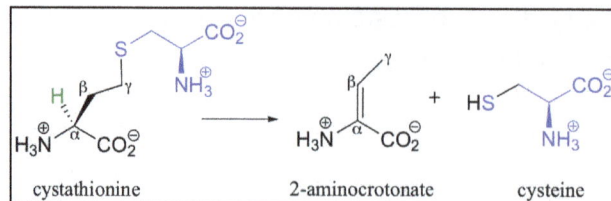

cystathionine                2-aminocrotonate                cysteine

Cystathionine first links to PLP in the normal way, using the amino group that is furthest away from the sulfur atom. In a familiar step, the alpha-proton is then abstracted by an enzymatic base, and the electron density is absorbed by PLP.

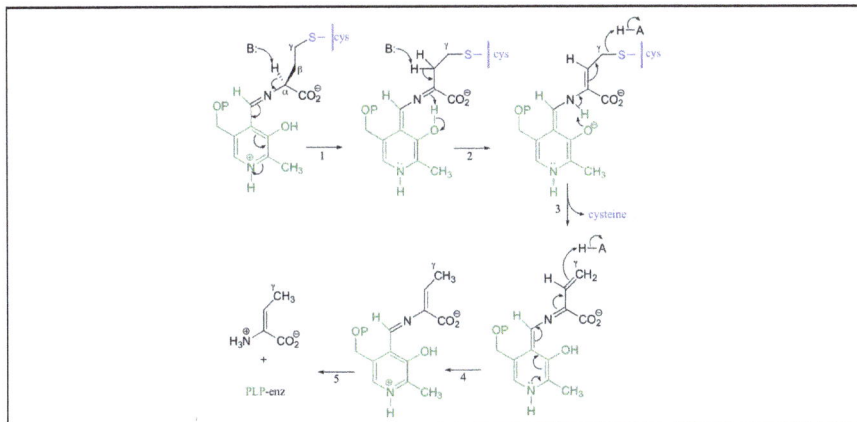

Next comes the new part - before anything happens to the electron density from the first proton abstraction, a second proton, this time from the beta-position on the side chain, is abstracted, forming an enamine intermediate This is made possible by the acidic phenolic proton on the pyridoxal ring of PLP. It is this second proton abstraction that is part of a recognizable elimination reaction, as the thiol group (which is actually a cysteine amino acid) leaves and a new pi-bond forms between $C\beta$ and $C\gamma$. This pi-bond is short-lived, however, as the electron density from the first proton abstraction, which has been 'hiding' in PLP all this time, finally bounces back and protonates $C\gamma$ With the usual imine transfer to an enzymatic lysine, the final product, 2-aminocrotonate, is released in.

Another related reaction is the PLP-dependant gamma-substitution, which starts off like a gamma-elimination, then reverses itself with the addition of a different nucleophile.

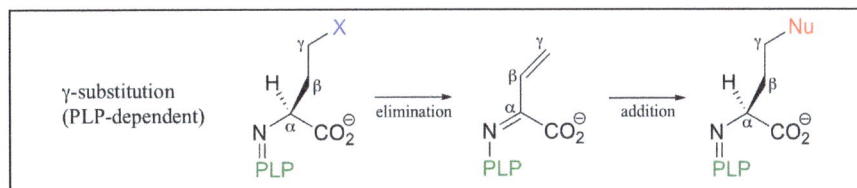

In the synthesis of methionine, cystathionine is obtained from O-succinyl homoserine and cysteine in a PLP-dependent gamma-substitution.

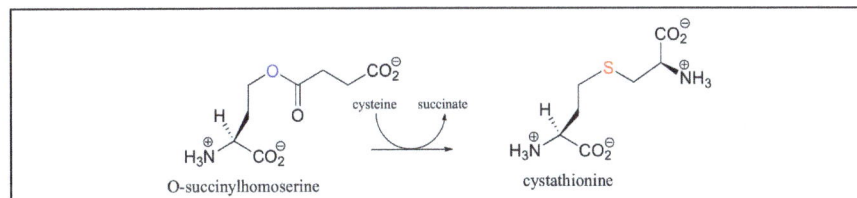

Altering the Course of a PLP Reaction Through Site-directed Mutagenesis.We have seen how PLP-dependent enzymes catalyze a group of reaction types - racemizations, retroaldols, transaminations, and eliminations - which, despite their apparent

diversity, are all characterized by a critical carbanion intermediate that is stabilized by
the electron sink of the PLP coenzyme. Given the similarities in the chemistry, it would
be reasonable to propose that the active site architecture of these enzymes might also
be quite close. This idea was nicely illustrated by an experiment in which site-directed
mutagenesis on a single active site amino acid of PLP-dependent alanine racemase was
sufficient to turn it into a retro-aldolase when provided with a suitable alternate substrate.

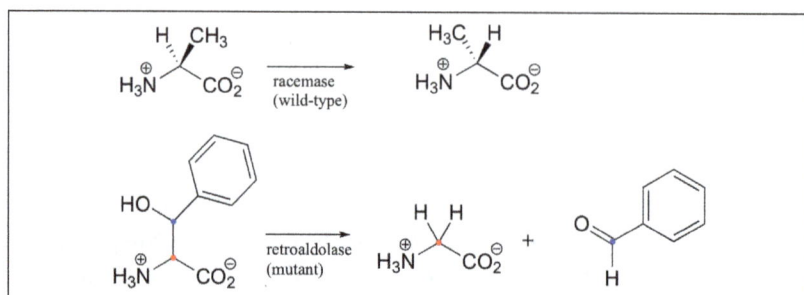

The catalytic base that abstracts the alpha-proton in the alanine racemase reaction is a
tyrosine, assisted by a nearby histidine. When researchers mutated the tyrosine to an
alanine, and substituted beta-hydroxytyrosine for the alanine substrate, a retro-aldol
reaction was catalyzed with remarkable efficiency.

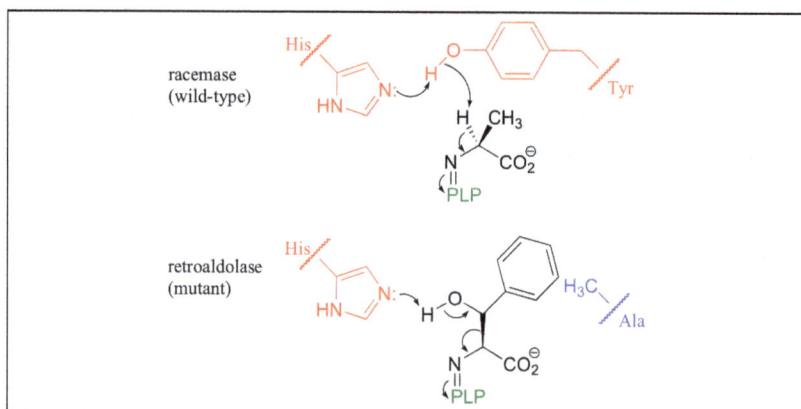

The basic histidine, with no tyrosine to deprotonate because of the mutation, instead
abstracts a proton from the beta-hydroxyl group of the new substrate, setting up a retroaldol cleavage. What the researchers did, essentially, was to swap a beta-hydroxy
amino acid substrate (capable of undergoing retroaldol cleavage) for the normal alanine substrate, then reposition the active site basic group so that a different acidic
proton was abstracted. That was all it took to change a racemase into a retroaldolase,
because the necessary electron sink system was all left in place. Why did they use the
unnatural amino acid beta-hydroxy tyrosine for the retro-aldol substrate rather than
threonine, which also has a beta-hydroxyl group and is closer in structure to alanine?
The researchers figured that the phenyl ring of beta-hydroxy tyrosine would fit nicely in
the space left empty due to the removal of the enzymatic tyrosine from the active site.

To reiterate: the point of this experiment was to demonstrate how similar the chemistry being catalyzed by different PLP-dependent enzymes really is - and, as a corollary, how similar the active sites are. Even though one would not normally consider a racemization to be closely related to a retroaldol cleavage, the mechanistic themes are very close, as evidenced by researchers' ability to change the reaction product by making a single active site mutation.

## References

- Active-site-of-an-enzyme: biologyreader.com, Retrieved 26 March, 2019

- Cofactor: biologydictionary.net, Retrieved 17 June, 2019

- Coenzyme: biologydictionary.net, Retrieved 9 January, 2019

- Hayashi H (2013). B Vitamins and Folate: Chemistry, Analysis, Function and Effects. Cambridge, UK: The Royal Society of Chemistry. p. 7. ISBN 978-1-84973-369-4

- Coenzyme-a, enzyme-reagents, enzyme-explorer, metabolomics, life-science: sigmaaldrich.com, Retrieved 19 April, 2019

- The-functions-of-coenzyme-a: livestrong.com, Retrieved 8 August, 2019

- Nicotinamide-adenine-dinucleotide, molecule-of-the-week: acs.org, Retrieved 23 July, 2019

- Nicotinamide-Adenine-Dinucleotide, Vitamins C-Cofactors-and-Coenzymes, Biological-Chemistry: libretexts.org, Retrieved 3 January, 2019

- Folate, vitamins: oregonstate.edu, Retrieved 13 March, 2019

- Biotin: newworldencyclopedia.org, Retrieved 6 May, 2019

- Pyridoxal-5-phosphate-p5p-highly-bioavailable-vitamin: medium.com, Retrieved 12 February, 2019

- Pyridoxal-phosphate---an-electron-sink-cofactor, Reactions-with-stabilized-carbanion-intermediates, Organic-Chemistry: chem.libretexts.org, Retrieved 21 May, 2019

# 4

# Fungal Enzymes

A majority of enzymes are produced by the fungi. They are integral as a source of biological diversity as well for the production of industrial enzyme products. Some of the important fungal enzymes are fungal ligninolytic enzymes, proteolytic enzyme, fungal phytases and fungal amylase. The chapter closely examines the key concepts of these fungal enzymes to provide an extensive understanding of the subject.

## Fungal Ligninolytic Enzymes

### Laccases

Laccases also known as benzenediol: oxygen oxidoreductase or p-diphenol oxidase belong to the oxidoreductase class.

Laccases are widely distributed in microorganisms, insects, and plants, showing a specific function in each of them. From this group, white rot fungi are the most studied laccases. These fungi use their enzymes to break the plant's lignocellulosic wall and obtain the host's nutrients. These fungal enzymes degrade the complex polyphenol structure that constitutes lignin, the principal recalcitrant component in the lignocellulosic wall. Besides their catalytic characteristics, these enzymes are attractive for biotechnological process, because they are extracellular and inducible, do not need a cofactor, and have low specificity. Laccase production is induced and increased by the addition of molecules as copper, dyes, or other recalcitrant compounds. Laccase employs oxygen as an oxidizing agent and cofactor instead of other expensive cofactors such as pantothenic acid, thiamine, or biotin.

Laccases have low substrate specificity; this characteristic allows the degradation of several compounds with a phenolic structure. Therefore, laccases have been employed in several areas such as bioremediation of aromatic recalcitrant compounds, treatment of effluents polluted with lignin, chemical synthesis, degradation of a wide number of textile dyes and biomass pretreatment for biofuel production.

### Laccase Structure and Catalytic Mechanism

Laccases have a primary structure of approximately 500 amino acid residues organized in three consecutive domains, with a Greek key β barrel topology. The three domains are distributed in a first domain with 150 initial amino acids, a second domain between the 150 and 300 residue, and a third domain from the 300 to 500 amino acid. The structure is stabilized by two disulfide bridges localized between domains I and II and between domains I and III. However, some laccases present three disulfide bridges. Melanocarpus albomyces has disulfide bridges inside domain I, another between domain I and domain III, and the last one between domain II and III.

Laccase belongs to multicopper oxidases (MCOs) and blue multicopper oxidases. Laccases have four copper (Cu) molecules in their active site which participate in oxygen reduction and water production. The laccases' four copper atoms are disseminated in three types of cores or places: type 1 Cu (T1), type 2 Cu (T2), and type 3 Cu (T3). These cores are in two metallic active sites: the mononuclear location T1 and the trinuclear location T2/T3. Type 1 Cu (T1) or blue Cu is a paramagnetic copper with a strong absorption at 600 nm (blue coloration). This signal is generated by the covalent union between Cu-Cys in the mononuclear location. In addition, T1 has the highest redox potential in the enzyme and participates in the enzymatic catalysis by oxidizing the substrate. Laccase's oxidative power is affected by the amino acids surrounding T1. Similar to T1, T2 is a paramagnetic cooper; however, it lacks absorption in the visible region of the spectrum. This Cu exhibits a strong absorption under electron paramagnetic resonance (EPR); technique employed to identify this spot. In addition, the T2/T3 core participates in the inhibition of the enzymatic activity by its interaction with anions such as fluoride or cyanide. On the other hand, T3 Cu is a binuclear core formed by a Cu (II) dimer. This site presents absorption at 330 nm and a lack of EPR signal. Type 3 coppers are diamagnetic and have a Cu-Cu union which participates in the compound oxidation as an electron acceptor.

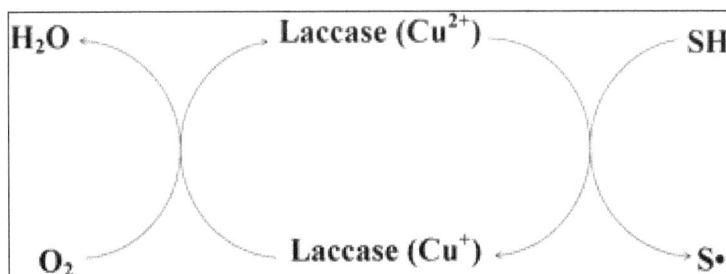

General diagram of laccase redox mechanism (SH reduced substrate, S• oxidized substrate).

In laccases, the different types of coppers have variations associated with the coordination between them and the amino acids in the active site. T1 Cu exhibits a different triangular planar coordination similar to other MCOs. Laccases display two histidines (His) and one cysteine (Cys) as equatorial ligands. Meanwhile, the other MCOs have an additional axial ligand with a methionine. In contrast, laccases did not have this axial extra bond, they have a leucine (Leu) or a phenylalanine (Phe) substituting methionine. On the other hand, T2 Cu exhibits coordination with two His and a water molecule; whereas, T3 Cu interacts with a water molecule and six His. In T3 Cu, the two Cu molecules share the water molecule and split the six His in two groups of three. The two T3

Cu interact with the oxygen atom; however, the interaction between water and the two Cu at the same time only happens when the enzyme is in its oxidized form.

Laccase crystalline structures evidenced that T2 Cu (EPR active) could be any of the three Cu in the trinuclear site. In this site, two of the Cu reduce and become silent to EPR. Meanwhile, the other one remains active to EPR. This interchange between T2 Cu and one of T3 coppers could be related to the reduction mechanism of the two oxygen molecules (cofactor), mechanism which is not clear for blue multicopper oxidases.

The enzymes' redox potential oscillates between 300 and 800 mV; these fluctuations depend on different factors such Cu ligands, distance among Cu, and the second sphere of amino acids in the active site. The last factor is associated with the degradation of recalcitrant compounds as lignin or industrial dyes.

Another structural characteristic in laccases is the presence of ligands: monosaccharides (mannose, N-acetyl glucosamine), ions like $Ca^{2+}$, and organic molecules like glycerol. The presence of these ligands produce significant differences between laccases, an example of that is the glycosylations exhibited in the isoenzymes from Trametes versicolor.

## Heme-peroxidases

After laccases, lignin peroxidase (LiP) and manganese peroxidase (MnP) are the most significant ligninolytic enzymes. These enzymes belong to the heme-proteins because they have the protoporphyrin IX as a prosthetic group. Similar to other heme-peroxidases such cytochrome c peroxidase and horseradish peroxidase, the catalytic cycle of LiP and MnP have three reactions.

Hydrogen peroxide causes the enzyme oxidation to produce the compound I and water:

$$\text{Reduced peroxidase} + H_2O_2 \rightarrow \text{Compound I} + H_2O$$

The modified enzyme (compound I) catalyzes the production of a free radical (S•) and a second modified form of the enzyme (compound II) by an electron transfer from the substrate (SH: reduced substrate):

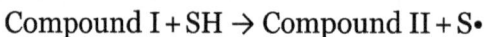

$$\text{Compound I} + SH \rightarrow \text{Compound II} + S•$$

Finally, compound II reacts with a second substrate molecule to produce another free radical and water; meanwhile, the enzyme reduces to its original form.

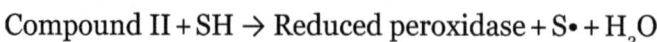

$$\text{Compound II} + SH \rightarrow \text{Reduced peroxidase} + S• + H_2O$$

## Lignin Peroxidase (LiP)

LiPs were originally discovered in nitrogen- and carbon-limited cultures of Phanerochaete chrysosporium. LiP possess high redox potential (700 to 1,400 mV), low optimum pH 3 to 4.5, and the ability to catalyze the degradation of a wide number of aromatic structures such veratryl alcohol (3,4-dimethoxybenzyl) and methoxybenzenes. LiP oxidizes aromatic rings moderately activated by electron donating substitutes; in contrast, common peroxidases participate in the catalysis of aromatic substrates highly activated (ammine, hydroxyl, etc.). An explanation for this type of catalysis is the production of veratryl alcohol radicals. These radicals have higher redox potential than LiP's compounds I and II and can participate in the degradation of compounds with high redox potential.

LiPs are monomeric glycosylated enzymes of 40 kDa, with 343 amino acids residues, 370 water molecules, a heme group, four carbohydrates, and two calcium ions. Their secondary structure is principally helicoidal. It contains eight major helixes, eight minor helixes, and two anti-parallel beta sheets. LiPs contain two domains at both sides of the heminic group. This group is inlaid in the protein but accessible to solvents via two small channels. The heminic cavity includes 40 residues, and it bonds to the protein via hydrogen bridges. Additionally, the heminic iron (Fe) coordinates with a His and a molecule of water. This His is associated with the high redox potential of LiP. The enzyme's redox potential rises when the His has a reduced imidazol character. In addition, a greater distance between the His and the heminic group increases the redox potential of the enzyme. This increment in the redox potential is a response to the electronic deficiency in the Fe of the porphyrin ring. In fact, this distance causes most differences among enzymes with similar porphyrin cores.

Another characteristic related with LiP's high redox potential is the invariant presence of a tryptophan residue (Trp171) in the enzymes' surface. Trp171 seems to facilitate electronic transference to the enzyme from substrates that cannot access into the heminic oxidative group. Additionally, Trp171 participates with the catalysis of veratryl alcohol, a metabolite produced by some ligninolytic fungi. Veratryl alcohol participates in the oxidation of different aromatic molecules. Some researchers conceptualized that this alcohol protects the enzyme from the action of $H_2O_2$ and participate as a redox mediator between the enzyme and substrates which cannot get inside the heminic center.

## Manganese Peroxidase (MnP)

Kuwahara in 1984 found the first MnP in batch cultures of Phanerochaete chrysosporium. They are glycoproteins with a molecular weight between 38 and 62.5 kDa , approximately 350 amino acid residues, and a 43% of identity with LiP sequence. MnP structure has two domains with the heminic group in the middle, ten major helixes, a minor helix, and five disulfide bridges. One of those bridges participates in the manganese (Mn) bonding site. This site is a characteristic that distinguishes MnP from other peroxidases.

The enzyme's catalytic cycle starts with the transference of two electrons from the heminic group to $H_2O_2$; it produces compound I and water. After that, compound I catalyzes the oxidation of one substrate molecule with the production of a free radical and compound II. Compound II oxidizes $Mn^{2+}$ to produce $Mn^{3+}$, the cation responsible to oxidize aromatic compounds. It is important to keep in mind that compound II demands $Mn^{2+}$ presence for its reaction; in contrast, compound I can oxidize $Mn^{2+}$ or the substrate. After $Mn^{3+}$ is stabilized by organic acids, it reacts non-specifically with organic molecules by removing an electron and a proton from the substrates. $Mn^{3+}$ is a small size compound with high redox potential, which diffuses easily in the lignified cell wall. Therefore, $Mn^{3+}$ starts the attack inside the plant cell wall which facilitates the penetration and action of the other enzymes.

## Sources of Ligninolytic Enzymes

The ligninolytic enzymes are a ubiquitous group of enzymes found in different types of organisms as plants, bacteria, insects, and fungi. In plants, laccases are the most documented ligninolytic enzyme; these are extracellular glycoproteins composed by a monomeric protein with a sugar component. Laccases have been found in different types of plants: the Japanese lacquer tree, mango, mung bean, peach, tobacco, zea mays, etc. In plants, laccases realize different kind of functions such as lignin synthesis, iron oxidation from Fe(II) to Fe(III), and regeneration of injured tissue. In plants, lignin synthesis is laccases' most significant function. In these organisms, they catalyze the monolignol dimerization, the basic molecules of lignin polymers and oligomers.

In bacteria, lignin breakdown is exhibited by three groups: actinomycetes, α-proteobacteria, and γ-proteobacteria. These three groups of bacteria were described in insects degrading wood microflora. These types of bacteria showed delignification in both in vitro and in vivo analyses. Streptomyces viridosporus, actinomycetes species, and Thermobifida fusca exhibited LiP activities. In contrast, Azospirillum lipoferum, Thermus thermophilus, Marinomonas mediterranea, Bacillus subtilis, and Streptomyces cyaneus showed laccase activity. Bacterial laccases are intracellular enzymes with a monomeric, multimeric, or homotrimeric structure without carbohydrate moiety. In nature, bacteria use laccases in spore protection and pigmentation. This type of enzymes exhibited resistance to high pH and temperatures.

In insects, the delignification process is done by ligninolytic enzymes produced by the insects or by the insects' microflora. The ligninolytic enzymes have been identified in different insects and different parts of them. An example of that are Nephotettix cincticeps (salivary glands), Manduca sexta (Malpighian tubules, midgut, fat body, and epidermis), Reticulitermes flavipes (gut), and Tribolium castaneum (cuticles). The principal functions of insect's ligninolytic enzymes are cuticle sclerotization and pigmentation, toxic compound oxidation, and polymerization reactions.

The fungal ligninolytic enzymes are the most well-known enzymes, and they occur in ascomycetes, basidiomycetes, and deuteromycetes. Inside these groups are several fungal species; however, the most studied fungi are: Trametes versicolor, Phanerochaete chrysosporium, Pleurotus ostreatus, Dichomitus squalens, Lentinula edodes, Irpex lacteus, and Cerrena maxima. The principal functions of fungal ligninolytic enzymes are lignin degradation, spore and fruiting body formation, and degradation of plant's toxic compounds.

## Biofuel Applications

In the last years, ligninolytic fungi and their enzymes have appeared as a new alternative for pretreatment processes. This technology has been useful principally in two processes, delignification and detoxification. In delignification, ligninolytic fungi or their enzymes have been used to reduce the lignin content in different types of feedstock. Whereas, in detoxification, ligninolytic enzymes have been used in the removal of chemical compounds from sugar hydrolysates after traditional pretreatments.

## Delignification

In delignification process, ligninolytic enzymes have been applied in four different methods:

1. Fungal delignification,

2. Enzymatic delignification,

3. Laccase-mediator system (LMS),

4. Integrated fungal fermentation (IFF).

Fungal delignification utilizes the complete ligninolytic microorganisms for biomass delignification. In those cases, the microorganisms grow with the target biomass in a submerged culture or a solid state fermentation. Both cultivations achieved good delignification percentages; however, these were not always related with high glucose yields. the lack of relation between high glucose yield and high delignification percentage occurs in different pretreatments and not only in microbial delignification. In fact, the principal issue with fungal pretreatment is the large duration of the process compared with the other pretreatment technologies. The length of time used by the microorganisms to obtain high delignification percentages is not less than 13 days and can be up to 40 or 50 days. However, this period will depend on the strain used. The effectiveness of microbial delignification has been improved by adding an alkali treatment previous the fungal pretreatment. The alkali pretreatment benefits were the reduction of the process duration and the increment in the glucose and ethanol yield.

Enzymatic delignification is the application of enzymatic extracts and purified or semi-purified ligninolytic enzymes to realize the lignin degradation. Enzymatic delignification employs commercial or native ligninolytic enzymes. Laccase is the most utilized enzyme followed by MnP and LiP; mixtures of two or three ligninolytic enzymes have been used for some delignification pretreatments. In those cases, the synergetic relationship between the ligninolytic enzymes improved biomass delignification. In wheat straw, ligninolytic enzymes generated a reduction in the cellulose conversion during the saccharification process. To avoid cellulases' inhibition, ligninolytic enzymes were deactivated before the saccharification process. Similar to fungal delignification, alkali pretreatments have been used in enzymatic delignification as an initial pretreatment to facilitate the delignification process. Enzymatic processes achieved similar delignification percentages than microbial pretreatment. However, enzymatic processes employ less amount of time (between 24 and 96 h) to produce the same delignification achieved by fungal pretreatment. Although this methodology reduces the delignification process duration, the current development of enzymatic delignification cannot compete with the conventional pretreatment technologies in terms of timespan and costs.

LMS utilize fungal laccases and redox mediators to achieve the delignification process. Redox mediators are chemical compounds, which act as electron carriers between the enzyme and the final substrate. Mediators' redox potential increases when the enzyme oxidizes them; after that, oxidized mediators react with the final substrate to recuperate the lost electrons. The increment in the redox potential allows the degradation of recalcitrant compounds such as lignin and different kinds of aromatic compounds that the enzyme by itself cannot break. The mediator compounds most used are 1-hydroxybenzotriazole (HBT), 2,2′-azino-bis(3-ethylbenzothiazoline-6-sulphonic acid) (ABTS), and some natural mediators as syringaldehyde or vanillin. The principal applications of laccase-mediator system are bleaching and delignification in the paper industry. However, this system has been used for delignification of some biofuel feedstocks. In cotton gin trash, the delignification was performed by a commercial LMS from Genencor. In that case, LMS was used after an ultrasonication and hot water pretreatments. LMS was improved by the addition of alkali-ultrasonication process before the hot water and LMS pretreatment.

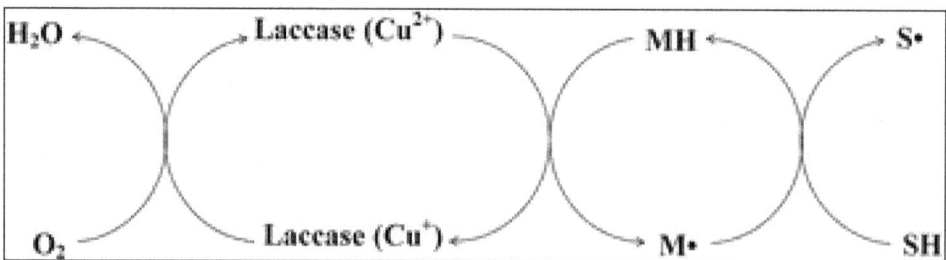

Laccase-mediator system's cycle

LMS pretreatment was utilized over corn stover by using HBT as mediator. In there, a correlation between laccase loading and enzymatic digestibility was found. Additionally, they found the importance of corn stover previous silage to increase LMS effectiveness. In wheat straw, LMS pretreatment was used after the steam explosion pretreatment and was performed before and simultaneously with the enzymatic saccharification. The results in both processes were similar. Thus, LMS can participate in a semi-consolidate process for delignification and saccharification. The application of LMS in elephant grass and eucalypt wood is the first report of LMS achieving significant delignification without a previous pretreatment. This improvement was achieved by using four steps of LMS pretreatment and NaOH washing. LMS is less used in biofuels than the paper industry. This lack of applications is related with the small development of specific mediators for feedstock delignification.

Integrated fungal fermentation (IFF) is a consolidated process where a fungus or a group of fungi transform biomass into ethanol without the participation of other treatments or microorganisms. Phlebia sp. MG-60 is a white rot fungus with the ability of selectively transforming lignin under solid state fermentation and produce ethanol from delignified biomasses under semi-aerobic submerged fermentation. Phlebia sp. MG-60 was selected from a group of other 12 fungi. From this group, Phlebia sp. MG-60 was the only one to exhibit the uncommon characteristic of degrade lignin and produce ethanol from cellulose. This fungal strain has been evaluated in the production of ethanol from different feedstocks such as hardwood kraft pulp, waste newspaper, sugarcane bagasse, and hard wood. The addition of basal media, organic compounds, and minerals to the culture media increased the ethanol yield and the delignification percentage. Phlebia sp. MG-60 produced an ethanol yield between 30% and 70% depending of the type of biomass utilized. The process duration is between 6 and 29 days and is associated with the type of biomass and the presence of pretreatment. The addition of alkali pretreatment generated a reduction in the process timespan from 21 days to 240 h. On the other hand, fungi co-culture is an alternative to generated and integrated process for ethanol production. In this case, ethanol is produced by mixing two types of fungi to develop delignification, saccharification, and fermentation. Coprinus comatus and Trametes reesei were cultivated to produce ethanol from corn stover. In this case, co-culture achieved greater delignification and cellulose conversion than monoculture.

Delignification using ligninolytic enzymes can be upgraded by improving the enzymatic catalysis using protein engineering. This area utilizes three types of approaches to modify ligninolytic enzymes: rational approaches, semi-rational approaches, and directed evolution. Rational approaches are the molecular modification (site direct mutation) of the ligninolytic enzyme sequence using prior structural information as foundation for the specific sequence modification. Rational approaches have been used to improve the laccase ability to degrade non-phenolic substrates and to enhance the capacity to oxidize bulky phenolic molecules. Semi-rational approaches employ saturation mutagenesis to modify hotspot residues in the enzymes. Saturation mutagenesis replaces the selected amino acids by all the codons that can generate the additional 19 amino acids. This approach exhibited the production of enzymes with three- to eightfold higher catalytic efficiencies. Directed molecular evolution is an approach that utilizes random mutation, gene recombination, and selection as fundaments. This molecular approach allows realizing the enzymatic design in absence of structural information. Direct molecular evolution has been used to improve catalytic activity and solvent tolerance.

The four delignification processes are compared in table. In enzymatic and microbial process, the pros observed are: positive environmental impact, high delignification and detoxification, low sugar losses, and the possibility of develop a consolidate process. On the other hand, microbial and enzymatic have different cons. The enzymatic processes cons are: high costs, low commercial availability, and mediator necessity. Meanwhile, in microbial processes, the principal limitation is the process duration. Besides the limitations and assets, ligninolytic fungi and their enzymes represent an interesting option to the conventional chemical and physical processes used in biofuel production.

Table: Delignification technique comparison

| Delignification process | Delignifi-cation | Sugar losses | Process duration | Economic | Environmental impact |
|---|---|---|---|---|---|
| Enzymatic | High | Low | 2 to 48 h | High costs | Positive |
| Fungal | High | Low | 6 to 45 days | Low costs | Positive |
| LMS | High | Low | 2 to 48 h | High costs | Depends of the mediator |
| IFF | Medium | Some | 1 to 12 weeks | Low costs | Positive |

## Detoxification

Some of the conventional biomass pretreatment technologies produce toxic compounds after treat biomass. These toxic compounds are classified in four groups:

1. Furan derivatives,

2. Pentose and hexose degradation,

3. Weak acids,

4. Phenolic compounds from lignin.

These compounds affect the fermentative microorganisms and the cellulolytic enzymes, which generate an overall ethanol yield reduction.

The detoxification process employs chemical, physical, and biological strategies. Chemical and physical techniques utilize technologies such as filtration, anion exchange chromatography, and NaOH precipitation; these techniques are expensive, produce sugar losses, and do not remove all the inhibitors in the hydrolysates. The biological detoxification process employs ligninolytic fungi or their enzymes to reduce the concentration of toxic compounds in feedstock hydrolysates. The detoxification process is performed after or before the saccharification process and has a duration between 1 to 12 h. Detoxification after or before saccharification is a controversial point. From the four types of toxic compounds produced in biomass hydrolysates, phenolic compounds are the most degraded type by ligninolytic enzymes. Phenolic compounds inhibited fermentation at concentration of 1 ppm. On the other hand, the other types of toxic compounds exhibit a variable behavior. These compound detoxification is affected by feedstock nature, enzyme characteristics, and pretreatment harshness. Enzymatic detoxification has been applied in the reduction of toxicity after different pretreatments such as steam explosion, strong acids, organosolv, and hot liquid water. In addition, ligninolytic enzyme detoxification increase fermentation rates and ethanol yield. As an example, in sugarcane bagasse, enzymatic detoxification generated an ethanol yield five times greater than ion exchange detoxification. Ligninolytic enzyme detoxification is a future option for biofuel industries. However, this enzymatic option needs to improve their ability to degrade other types of toxic compounds from hydrolysates.

# Proteolytic Enzyme

Proteolytic enzyme, also called protease, proteinase, or peptidase is any of a group of enzymes that break the long chainlike molecules of proteins into shorter fragments (peptides) and eventually into their components, amino acids. Proteolytic enzymes are present in bacteria, archaea, certain types of algae, some viruses, and plants; they are most abundant, however, in animals.

There are different types of proteolytic enzymes, which are classified according to sites at which they catalyze the cleavage of proteins. The two major groups are the exopeptidases, which target the terminal ends of proteins, and the endopeptidases, which target sites within proteins. Endopeptidases employ various catalytic mechanisms; within this group are the aspartic endopeptidases, cysteine endopeptidases, glutamic endopeptidases, metalloendopeptidases, serine endopeptidases, and threonine endopeptidases. The term oligopeptidase is reserved for those enzymes that act specifically on peptides.

Among the best-known proteolytic enzymes are those that reside in the digestive tract. In the stomach, protein materials are attacked initially by a gastric endopeptidase known as pepsin. When the protein material is passed to the small intestine, proteins, which are only partially digested in the stomach, are further attacked by proteolytic enzymes secreted by the pancreas. These enzymes are liberated in the small intestine from inactive precursors produced by the acinar cells in the pancreas. The precursors are called trypsinogen, chymotrypsinogen, proelastase, and procarboxypeptidase. Trypsinogen is transformed to an endopeptidase called trypsin by an enzyme (enterokinase) secreted from the walls of the small intestine. Trypsin then activates the precursors of chymotrypsin, elastase, and carboxypeptidase. When the pancreatic enzymes become activated in the intestine, they convert proteins into free amino acids, which are easily absorbed by the cells of the intestinal wall. The pancreas also produces a protein that inhibits trypsin.

# Fungal Phytases

Fungal phytases are histidine acid phosphatases, a subclass of acid phosphatases, which catalyse the hydrolysis of phytic acid resulting in the release of phosphate moieties and thus mitigate its antinutritional properties. The supplementation of feed with phytases increases the bioavailability of phosphorus and minerals in non-ruminant animals and reduces the phosphorus pollution due to phosphorus excretion in the areas of intensive livestock production. Although phytases are reported in plants, animals and micro-organisms, fungal sources are used extensively for the production of phytases on a commercial scale. Phytases have been produced by fungi in both solid-state fermentation (SSF) and submerged fermentation (SmF). The fungal phytases are high molecular weight proteins ranging from 35 to 500 kDa. They are optimally active within pH and temperature ranges between 4.5 and 6.0, and 45 and 70 °C respectively. Phytate degradation leads to amelioration in the nutritional status of foods and feeds by improving the availability of minerals, phosphorus and proteins in non-ruminant animals and human beings and thus mitigates the environmental phosphorus pollution.

# Fungal Amylase

Fungal Amylase is an alpha amylase enzyme preparation produced by Aspergillus oryzae. It is available as a liquid or as a powder formulation. This rapid acting hydrolase is active throughout the acidic pH range, through the neutral pH range and well into the mildly alkaline range. It readily degrades a wide variety of starch-containing substrates. This product is not extremely heat-resistant compared with enzymes from bacterial sources but exhibits adequate heat resistance for many industrial and most agricultural applications where its pH range is often complementary to alpha amylase from bacterial sources.

Fungal Amylase is used in commercial starch modification, cleaning compounds, textile processing, fermentation pretreatment, and animal feed supplements. It can supplement the deficiency of endogenous amylase of young animals. It can also increase the utilization of feedstuffs by hydrolyzing the starch contained in feed to generate dextrins and sugars. In addition, this enzyme facilitates the contact between the digestive enzymes and chyme to improve nutrient absorption and utilization. Fungal amylase lowers feed conversion ratio (FCR) and improves animal production performance and utilization of energy.

## Classification of Amylase

1. α-Amylase: The-amylases are calcium metalloenzymes, completely unable to function in the absence of calcium. By acting at random locations along the starch chain, α-amylase breaks down long-chain carbohydrates, ultimately yielding maltotriose and maltose from amylose, or maltose, glucose and "limit dextrin" from amylopectin. Because it can act anywhere on the substrate, α-amylase tends to be faster-acting than-amylase. In animals, it is a major digestive enzyme and its optimum pH is 6.7-7.0. In human physiology, both the saliva and pancreatic amylases are-amylases. They are discussed in much more detail at alpha-Amylase. Also found in plants (adequately), fungi (ascomycetes and basidiomycetes) and bacteria (Bacillus).

Alpha amylases are one of the important and widely used enzymes whose spectrum of applications has widened in many sectors such as clinical, medicinal and analytical chemistry. Besides their use in starch saccaharification they also find applications in food, baking, brewing, detergent, textile, paper and distilling industry. Alpha amylases (endo-1,-D-glucan glucohydrolase) constitute the family of endo amylases that randomly cleave the-D- glycosidic linkages between adjacent glucose units in the linear amylose chain with retention of α-anomeric configuration of the products. Production of fungal alpha amylases has been investigated through submerged (SMF) and solid state fermentation (SSF).

2. β-Amylase: Another form of amylase, β-amylase is also synthesized by bacteria, fungi and plants. Working from the non-reducing end, β-amylase catalyzes the hydrolysis of the second-1,4 glycosidic bond, cleaving off two glucose units (maltose) at a time. During the ripening of fruit, β-amylase breaks starch into maltose, resulting in the sweet flavor of ripe fruit.

Both α-amylase and β-amylase are present in seeds; β-amylase is present in an inactive form prior to germination, whereas-amylase and proteases appear once germination has begun. Cereal grain amylase is key to the production of malt. Many microbes also produce amylase to degrade extracellular starches. Animal tissues do not contain β-amylase, although it may be present in microorganisms contained within the digestive tract.

3. γ-Amylase: In addition to cleaving the last α(1-4)glycosidic linkages at the nonreducing end of amylose and amylopectin, yielding glucose,-amylase will cleave α(1-6) glycosidic linkages. Unlike the other forms of amylase, γ-amylase is most efficient in acidic environments and has an optimum pH of 3.

4. Fungal Amylases: Bacteria and fungi secrete amylase to the outside of the cells to carry out extra-cellular digestion.

Fungal amylases are used for hydrolyzing carbohydrate, protein and other constitutes of soybeans, wheat into peptides, amino acid, sugars and other low molecular weight compounds. The amylase producing strains of Aspergillus niger have spore bearing heads which are large, tightly packed, globular and may be black or brownish black. They are considered to be mesophilic with optimal temperature for growth between 25 °C and 35 °C. They are aerobic in nature and can grow over a wide range of hydrogen ion concentration. These organisms can utilize different kinds of agricultural wastes from simple to complex ones, which make them easy to cultivate and maintain in the laboratory.

Filamentous fungi are microorganisms secrete large amounts of protein in culture medium. Aspergillus niger has been described as secreting an α-amylase and glucoamylase of a number of different molecular weights in submerged culture. Filamentous fungi have been used for the industrial production of a wide variety of native products, such as antibiotic (e.g., penicillin and cephalosporin), organic acid (Citric and acetic acid) and commercial enzymes (e.g., protease, catalase, amylase). Aspergillus niger is an acidifient mould thanks to its important hydrolytic capacities in the α–amylase production and its tolerance of acidity (pH < 3), it allows the avoidance of bacterial contamination.

The majority of industrial enzymes being used currently belong to the hydrolase group, which utilizes several different natural substrates. Detergent industries are the primary consumers of enzymes, in terms of both volume and value. For years microorganisms have been the principal source of many different enzymes, which were identified after research and currently find their main uses in industrial application several industries employ microbial amylolytic enzymes and a growing industrial application for them is the enzymatic conversion of starch into variants of sugar solution. It has been reported that solid state fermentation is the most appropriate process in developing countries due to the advantages it offers. The hyphal mode of fungal growth and their good tolerance to low water activity and high osmotic pressure conditions make fungi efficient and competitive in natural microflora for bioconvension of solid substrates.

α-Amylase from Aspergillus oryzae was the first microbial enzyme to be manufactured for sale and was named by solid state cultivation for many years. The manufacture had switched to submerged fermentation and such methods have been reviewed. Some difficulties were encountered in making this change since the most effective preparation of some applications contains other enzymes, especially amyloglycosidases and the submerged methods gives a narrower spectrum of additional additives. So, it is worthwhile to isolate a suitable strain of Aspergillus niger for efficient mechanism selection of suitable production media is very essential for growth of microorganisms as well as production of enzyme. The production of alpha-amylase by mouth has been greatly affected by the cultural and nutritional requirement.

In most developing countries of the tropics carbohydrate based agricultural products like starchy tubers and cereal occur abundantly. Starchy tubers such as cassava, yam, sweet potato and calcium are important staple foods in the diet of people in most developing countries of the tropics. In these countries, they are widely distributed and more cultivated than cereal. However, despite their importance, a large proportion of their tubers are cost yearly due to inadequate and ineffective storage facilities. Application of theseagro-industrial residues in bioprocesses also solve pollution problems, which their disposal may otherwise cause with the advent of biotechnological innovations, mainly in the area of enzyme and fermentation technology, many new areas have opened for their utilization as raw materials for the production of value added fine products 29.

## Biochemical Properties of Amylases

1. Substrate Specificity: As holds true for the other enzymes, the substrate specificity of -amylase varies from microorganism to a microorganism. In general, α-amylases display highest specificity towards starch followed by amylose, amylopectin, cyclodextrin, glycogen and maltotriose.

2. pH Optima and Stability: The pH optima of α–amylases vary from 2 to 12. α-Amylases from most bacteria and fungi have pH optima in the acidic to neutral range. -Amylase from Alicyclobacillus acidocaldarius showed an acidic pH optimum of 3, in contrast to the alkaline amylase with optima of pH 9 to 10.5 reported from an alkalophilic Bacillus sp.. Extremely alkalophilic α-amylase with pH optima of 11 to 12 has been reported from Bacillus sp. In some cases, the pH optimum was observed to be dependent upon temperature as in the case of Bacillus stearothermophilus and on calcium as in the case of Bacillus stearothermophilus. α –Amylases are generally stable over a wide range of pH from 4 to 11, however, -amylases with stability in a narrow range have also been reported.

3. Temperature Optima and Stability: The temperature optimum for the activity of amylase is related to the growth of the microorganism. The lowest temperature optimum is reported to be 25 to 30 °C for Fusarium oxysporum amylase and the highest of 100 and 130°C from archaebacteria, Pyrococcus furiosus and Pyrococcus woesei, respectively. Temperature optima of enzymes from Micrococcus varians are calcium dependent and that from H. meridiana is sodium chloride dependent. Thermostabilities have not been estimated defector in many studies. Thermostabilities as high as 4 hours at 100°C have been reported for Bacillus licheniformis. Many factors affect thermostability. These include the presence of calcium, substrate and other stabilizers. The stabilizing effect of starch was observed in amylases from Bacillus licheniformis, Lipomyces kononenkoae and Bacillus sp. Thermal stabilization of the enzyme in the presence of calcium has also been reported from time to time.

4. Molecular Weight: Molecular weights of amylases vary from about 10 to 210 kDa. The lowest value, 10 kDa for Bacillus caldolyticus and the highest of 210 kDa for Chloroflexus aurantiacus has been reported. Molecular weights of microbial amylases are usually 50 to 60 kDa as shown directly by analysis of cloned amylase genes and deduced amino

acid sequences. Carbohydrate moieties raise the molecular weight of some amylases. Glycoproteins have been detected in Aspergillus oryzae, L. kononenkoae, Bacillus stearothermophilus and Bacillus subtilis strains. Glycosylation of bacterial proteins is rare. A carbohydrate content as high as 56% has been reported in S. castelii whereas this is about 10% for other amylases.

5. Inhibitors: Many metal cations, especially heavy metal ions, sulphydryl group reagents, N-bromosuccinimide, hydroxyl mercuribenzoic acid, iodoacetate, BSA, EDTA and EGTA inhibit amylases.

6. Calcium and Stability: Amylase is a metalloenzyme, which contains at least one $Ca^{2+}$ ion. The affinity of $Ca^{2+}$ to amylase is much stronger than that of other ions. The amount of bound calcium varies from one to ten. Crystalline Taka- amylase A (TAA) contains ten $Ca^{2+}$ ions but only one is tightly bound. Calcium free enzymes can be reactivated by adding $Ca^{2+}$ ions. Some studies have been carried out on the ability of other ions to replace $Ca^{2+}$ as $Sr^{2+}$ in Bacillus caldolyticus amylase. $Ca^{2+}$ in TAA has been substituted by $Sr^{2+}$ and $Mg^{2+}$ in successive crystallization in the absence of $Ca^{2+}$ and in excess of $Sr^{2+}$ and $Mg^{2+}$. EDTA inactivated TAA can be reactivated by $Sr^{2+}$, $Mg^{2+}$ and $Ba^{2+}$. In the presence of $Ca^{2+}$, amylases are much more thermostable than without it. Amylase from Aspergillus oryzae is inactivated in the presence of $Ca^{2+}$, but retains activity after EDTA treatment. There are also reports where $Ca^{2+}$ did not have any effect on the enzyme.

7. Fermentation Studies on Amylase Production: Solid state fermentation compared to submerged fermentation is more simple, require lower capital has superior productivity, reduced energy requirement, simpler fermentation media and absence of rigorous control of fermentation parameters, uses less water and produces lower waste water, has easier control of bacterial contamination and requires low cost of downstream processing. Industrially important enzymes have traditionally been obtained from submerged fermentation because of the ease of handling and greater control of environmental factors such as temperature and pH. However, solid-state fermentation constitutes an interesting alternative since the metabolites so produced are less costly.

The effect of environmental conditions on the regulation of extracellular enzymes in batch cultures is well documented. A lot of work on the morphology and physiology of α -amylase production by Aspergillus oryzae during batch cultivation has been done. Accordingly, morphology of Aspergillus oryzae was critically affected by the growth pH. In a series of batch experiments, the authors observed that at pH 3.0 to 3.5, freely dispersed hyphal elements were formed. In the pH range 4 to 5, both pellets and freely dispersed hyphal fragments were observed whereas at pH higher than 6 pellets were the only growth forms recorded. Other groups have recorded similar observations for other strains of Aspergillus oryzae. The optimum growth temperature was found to be 35 °C. It is demonstrated that when glucose was exhausted the biomass production stopped whereas the secretion of α -amylase increased rapidly.

A decline in enzyme production was also accompanied by morphological and metabolic variations during continuous cultivation. The industrial exploitation of SSF for enzyme production has been confined to processes involving fungi and it is generally believed that these techniques are not suitable for bacterial cultivation. The use of the SSF technique in α –amylase production and its specific advantages over other methods has been discussed extensively. In the solid state fermentation process, the solid substrate not only supplies the nutrients to the culture, but also serves as an anchorage for the microbial cells. The moisture content of the medium changes during fermentation as a result of evaporation and metabolic activities and thus the optimum moisture level of the substrate is therefore most important.

## Uses of Amylase

Amylases are among the most important hydrolytic enzymes for all starch based industries and the commercialization of amylases is oldest with first use in 1984, as a pharmaceutical aid for the treatment of digestive disorders. In the present day scenario, amylases find application in all the industrial processes such as in food, detergents, textiles and in paper industry, for the hydrolysis of starch. In this light, microbial amylases have completely replaced chemical hydrolysis in the starch processing industry. They can also be of potential use in the pharmaceutical and fine chemical industries. Today, amylases have the major world market share of enzymes. Several different amylase preparations are available with various enzyme manufacturers for specific use in varied industries.

Amylase enzymes find use in bread making and to break down complex sugars such as starch (found in flour) into simple sugars. Yeast then feeds on these simple sugars and converts it into the waste products of alcohol and $CO_2$. This imparts flavour and causes the bread to rise. While Amylase enzymes are found naturally in yeast cells, it takes time for the yeast to produce enough of these enzymes to break down significant quantities of starch in the bread. This is the reason for long fermented doughs such as sourdough. Modern bread making techniques has included amylase enzymes (often in the form of malted barley) into bread improver thereby making the bread making process faster and more practical for commercial use. When used as a food additive Amylase has E number E1100 and may be derived from swine pancreas or mould mushroom.

Amylase is also used in clothing and dishwasher detergents to dissolve the starches from fabrics and dishes. Workers in factories that work with amylase for any of the above uses are at increased risk of occupational asthma. 5-9% of bakers have a positive skin test and a fourth to a third of bakers with breathing problems are hypersensitive to amylase. An inhibitor of alpha-amylase called phaseolamin has been tested as a potential diet aid.

Blood serum amylase may be measured for purposes of medical diagnosis. A normal concentration is in the range 21-101 U/L. A higher than normal concentration may reflect one of several medical conditions, including acute inflammation of the pancreas (concurrently with the more specific lipase), but also perforated peptic ulcer, torsion of an ovarian cyst, strangulation isles, macroamylasemia and mumps. Amylase may be measured in other body fluids, including urine and peritoneal fluid.

In molecular biology, the presence of amylase can serve as an additional method of selecting for successful integration of a reporter construct in addition to antibiotic resistance. As reporter genes are flanked by homologous regions of the structural gene for amylase, successful integration will disrupt the amylase gene and prevent starch degradation, which is easily detectable through iodine staining.

# Lignocellulosic Residues: Biodegradation and Bioconversion by Fungi

Lignocellulose is the major component of biomass, comprising around half of the plant matter produced by photosynthesis (also called photomass) and representing the most abundant renewable organic resource in soil. It consists of three types of polymers, cellulose, hemicellulose and lignin that are strongly intermeshed and chemically bonded by non-covalent forces and by covalent cross linkages. Only a small amount of the cellulose, hemicellulose and lignin produced as by-products in agriculture or forestry is used, the rest being considered waste. Many microorganisms are capable of degrading and utilizing cellulose and hemicellulose as carbon and energy sources. However, a much smaller group of filamentous fungi has evolved with the ability to break down lignin, the most recalcitrant component of plant cell walls. These are known as white-rot fungi, which possess the unique ability of efficiently degrading lignin to $CO_2$. Other lignocellulose degrading fungi are brown-rot fungi that rapidly depolymerize cellulosic materials while only modifying lignin. Collectively, these wood and litter-degrading fungi play an important role in the carbon cycle. In addition to lignin, white-rot fungi are able to degrade a variety of persistent environmental pollutants, such as chlorinated aromatic compounds, heterocyclic aromatic hydrocarbons, various dyes and synthetic high polymers.

This degradative ability of white-rot fungi is due to the strong oxidative activity and low substrate specificity of their ligninolytic enzymes. Little is known about the degradation mechanisms of lignocellulose by soft rot fungi, in contrast to white and brown rot fungi. it is clear that some soft-rot fungi can degrade lignin, because they erode the secondary cell wall and decrease the content of acid-insoluble material (Klason lignin) in angiosperm wood. Soft rot fungi typically attack higher moisture, and lower lignin content materials. The genome sequences from different fungi such as; Phanerochaete chrysosporium strain RP8, Coprinopsis cinerea, Postia placenta, Pleurotus ostreatus, Agaricus bisporus, Schizophyllum commune have been revealed and its genomic information may greatly facilitate our understanding of the lignocellulose biodegradation process. World-wide lignocellulosic residue generation every year results in pollution of the environment and in loss of valuable materials that can be bioconverted to several added-value products. Lignin can be removed by chemical or physical pre-treatment which then permits efficient bioconversion. Pre-treatment can also be carried out microbiologically. This has the advantages over non-biological procedures of producing potentially useful by-products and minimal waste. This review will focus on the use of fungi in the biodegradation of lignocellulose, aspects of bioconversion and worldwide lignocellulosic residues.

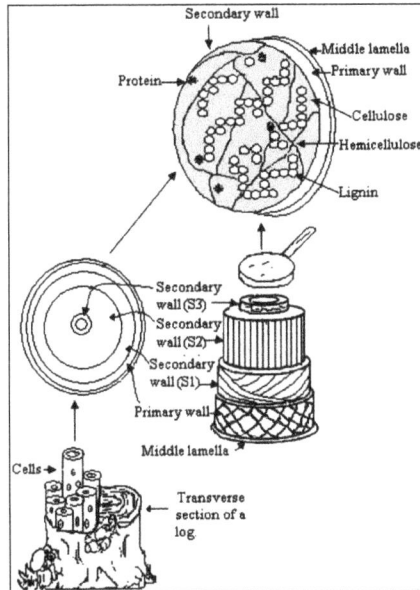

Composition of lignocellulosic residues. Cellulose , hemicellulose and lignin.

## Composition of Lignocellulosic Residues

The major component of lignocellulosic materials is cellulose, followed by hemicellulose and lignin. Cellulose and hemicellulose are macromolecules constructed from different sugars; whereas lignin is an aromatic polymer synthesized from phenylpropanoid precursors. The composition and proportions of these compounds vary between plants. Cellulose is a linear polymer that is composed of D-glucose subunits linked by β-1,4 glycosidic bonds forming the dimer cellobiose. These form long chains (or elemental fibrils) linked together by hydrogen bonds and van der Waals forces. Cellulose usually is present as a crystalline form and a small amount of nonorganized cellulose chains forms amorphous cellulose. In the latter conformation, cellulose is more susceptible to enzymatic degradation. Cellulose appears in nature to be associated with other plant compounds and this association may affect its biodegradation. Hemicellulose is a polysaccharide with a lower molecular weight than cellulose. It is formed from D-xylose, D-mannose, Dgalactose, D-glucose, L-arabinose, 4-O-methyl-glucuronic, D-galacturonic and D-glucuronic acids. Sugars are linked together by β-1,4- and sometimes by β-1,3-glycosidic bonds. The main difference between cellulose and hemicellulose is that hemicellulose has branches with short lateral chains consisting of different sugars and cellulose consists of easily hydrolyzable oligomers.

Lignin is linked to both hemicellulose and cellulose, forming a physical seal that is an impenetrable barrier in the plant cell wall. It is present in the cellular wall to give structural support, impermeability and resistance against microbial attack and oxidative stress. It is an amorphous heteropolymer, non-water soluble and optically inactive that is formed from phenylpropane units joined together by non-hydrolyzable linkages.

This polymer is synthesized by the generation of free radicals, which are released in the peroxidase-mediated dehydrogenation of three phenyl propionic alcohols: coniferyl alcohol (guaiacyl propanol), coumaryl alcohol (p-hydroxyphenyl propanol), and sinapyl alcohol (syringyl propanol). This heterogeneous structure is linked by C–C and aryl-ether linkages, with aryl-glycerol β-aryl ether being the predominant structures.

## Biodegradation of Lignocellulosic Residues

The organisms predominantly responsible for lignocellulose degradation are fungi, and the most rapid degraders in this group are basidiomycetes. The ability to degrade lignocellulose efficiently is thought to be associated with a mycelial growth habit that allows the fungus to transport scarce nutrients such as nitrogen and iron, to a distance into the nutrient-poor lignocellulosic substrate that constitutes its carbon source. The fungal degradation occurs exocellularly, either in association with the outer cell envelope layer or extracellularly, because of the insolubility of lignin, cellulose and hemicellulose. Fungi have two types of extracellular enzymatic systems: the hydrolytic system, which produces hydrolases that are responsible for polysaccharide degradation; and a unique oxidative and extracellular ligninolytic system, which degrades lignin and opens phenyl rings. Several microorganisms, mainly fungi, have been isolated and identified as lignocellulolytic organisms. The most widely studied white-rot organism is P.chrysosporium, which is one of the holobasidiomycetes. Trichoderma reesei and its mutants are the most studied ascomycete fungi, and is used for the commercial production of hemicellulases and cellulases. Not even whiterot fungi are known to be capable of using lignin as a sole carbon and energy source, and it is generally believed that lignin break down is necessary to gain access to cellulose and hemicellulose. Although white-rot basidiomycetes have been shown to efficiently mineralize lignin, species differ gross morphological patterns of decay they cause. P.chrysosporium strains simultaneously degrade cellulose, hemicellulose and lignin, whereas others such as Ceriporiopsis subvermispora tend to remove lignin in advance of cellulose and hemicellulose. Brown rot mechanism has likely evolved independently multiple times from white rot decay fungi. Presumably, because lignin breakdown is energetically unfavourable, selection has favoured a mechanism which can specifically attack the cellulose and hemicellulose components.

## Lignin Biodegradation

Lignin biodegradation by white-rot fungi is an oxidative process and phenol oxidases are the key enzymes. Of these, lignin peroxidases (LiP), manganese peroxidases (MnP) and laccases  from white-rot fungi (especially Botrytis cinerea, P. chrysosporium, Stropharia coronilla, P. ostreatus and Trametes versicolor) have been studied. LiP and MnP oxidize the substrate by two consecutive one-electron oxidation steps with intermediate cation radical formation. LiP and MnP were discovered in the mid-1980s in P. chrysosporium and described as true lignases because of their high potential redox value. LiP degrades non-phenolic lignin units (up to 90% of the polymer), whereas MnP generates $Mn^{3+}$, which acts as a diffusible oxidizer on phenolic or non-phenolic lignin units via lipid peroxidation reactions. Laccase are blue copper oxidases that catalyze the one-electron oxidation of phenolics and other electron-rich substrates. Recently, other enzymes involved in lignin degradation have been reported. These include aryl- alcohol oxidase (AAO) described in Pleurotus eryngii  and other fungi, and P. chrysosporium glyoxal oxidase. Fungal aryl-alcohol dehydrogenases (AAD) and quinone reductases (QR) are also involved in lignin degradation.

Laccases or ligninolytic peroxidases (LiP and MnP) produced by white-rot fungi oxidize the lignin polymer, thereby generating aromatic radicals (a). These evolve in different non-enzymatic reactions, including C-4-ether breakdown (b), aromatic ring cleavage (c), Cα–Cβ breakdown (d), and demethoxylation (e). The aromatic aldehydes released from Cα–Cβ breakdown of lignin, or synthesized de novo by the fungus (f, g), are the substrates for $H_2O_2$ generation by AAO in cyclic redox reactions also involving AAD. Phenoxy radicals from C4-ether breakdown  (b) can repolymerize on the lignin polymer (h) if they are not first reduced by oxidases to phenolic compounds (i). The phenolic compounds formed can be again reoxidized by laccases or peroxidases (j). Phenoxy radicals can also be subjected to Cα–Cβ breakdown (k), yielding p-quinones. Quinones from g and/or k contribute to oxygen activation in redox cycling reactions involving oxygen activation in redox cycling reactions with QR, laccases, and peroxidases (l, m). This results in reduction of the ferric iron present in wood (n), either by superoxide cation radical or directly by the semiquinone radicals, and its reoxidation with concomitant reduction of $H_2O_2$ to a hydroxyl free radical (OH×) (o).

The latter is a very mobile and very strong oxidizer that can initiate the attack on lignin (p) in the initial stages of wood decay, when the small size of pores in the still-intact cell wall prevents the penetration of ligninolytic enzymes. Then, lignin degradation  proceeds by oxidative attack of  the enzymes described above. In the final steps, simple products from lignin degradation enter the fungal hyphae and are incorporated into intracellular catabolic routes. Fungal feruloyl and p-coumaroyl esterases are capable of releasing feruloyl and p-coumaroyl units and play an important role in biodegradation of recalcitrant cell walls in grasses. These enzymes act synergistically with xylanases to disrupt the hemicellulose-lignin association, without mineralization of lignin per se. Therefore, hemicellulose degradation is required before efficient lignin removal can com- mence. In P. chrysosporium, a co-metabolizable carbon source is essential for lignin degradation, and it is produced in response to nitrogen starvation. This indicates that the ligninolytic system is formed as part of secondary metabolism in this organism. Carbohydrate starvation likewise leads to a rapid but transient onset of ligninolytic activity. Elevated oxygen levels increase the rate of lignin biodegradation through the production of hydrogen peroxide as the extracellular oxidant and the subsequent induction of ligninolytic activity. The hydrogen peroxide is derived from the co-metabolism of cellulose and hemicellulose.

## Cellulose Biodegradation

Cellulolytic microorganisms can establish synergistic relationships with non-cellulolytic species in cellulosic wastes; the interaction leads to complete degradation of cellulose. Microorganisms capable of de- grading cellulose produce a battery of enzymes with different spec- ificities, working together. Cellulases responsible for the hydrolysis of cellulose, are composed of a complex mixture of enzyme proteins with different specificities to hydrolyze the β-1,4-glycosidic linkages bonds. Cellulases can be divided into three major enzyme activity classes. These are endogluca- nases or endo-1-4-β-glucanase, cellobiohydrolase and β-glucosidase. Endoglucanases, often called carboxymethylcellulases (because of the artificial substrate used for their detection), are thought to initiate attack randomly at multiple internal sites in the amorphous regions of the cellulose fibre which opens-up sites for subsequent attack by the cellobio-hydrolases. Cellobiohydrolase, often called exoglucanase, is the major component of the fungal cellulase system accounting for 40–70% of the total cellulase proteins, and can hydrolyze highly crystalline cel-Cellobiohydrolases remove monomers and dimers from the end of the glucan chain. β- glucosidase hydrolyzes glucose dimers and in some cases cellulose- oligosaccharides to glucose. Generally, the endoglucanases and have multiple distinct variants of endo- and exo-glucanases.

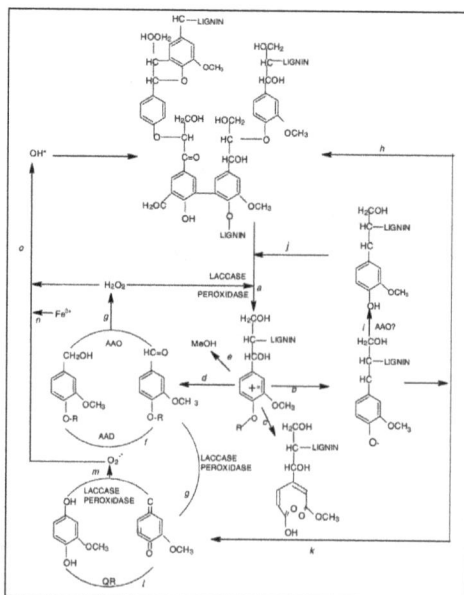

Lignin biodegradation process by white rot fungi

## Hemicellulose Biodegradation

Although similar enzymes are involved for cellulose and hemi- cellulose biodegradation, more enzymes are required for the latter's complete degradation because of its greater heterogeneity compared with cellulose. Hemicelluloses are biodegraded to monomeric sugars and acetic acid. Xylan is the main carbohydrate found in hemicellulose. Its complete degradation requires the cooperative action of a variety of hydrolytic enzymes. Hemicellulases are frequently classified according to their action on distinct substrates, endo-1,4-β-xylanase generates oligosaccharides from the cleavage of xylan and xylan 1,4-β- xylosidase produces xylose from oligosaccharides.

Table: Fungal cellulases with highest specificity activity.

| Enzyme | Organism | Substrates | Specific activity ($\mu$mol min$^{-1}$ mg$^{-1}$) | Opt. T (°C) | Opt. pH |
|---|---|---|---|---|---|
| Mannnan endo-1,4-β-mannosidase | Sclerotium rolfsii | Galactoglucomannan/ galactomannans/ glucomannan/ mannnans | 475 | 72-74 | 3.3 |
| Cellulase | Aspergillus níger | Carboxymethylcellulose/ cellohexaose/ cellopentaose/ cellotetraose/cellotriose/ cellulose | 194 | 70 | 5 |
| 1,3-βglucan glucohydrolase | Achlya bisexuals | Glucan/laminarin/ neutral glucan/ phosphoglucan | 7840 | 30 | 6 |
| 1,3-1,4-β-dglu-can glucanohydro-lase | Orpinomy-ces sp. | β-d-glucan/lichenin | 3659 | 45 | 5.8 |
| 1,3-β-dglucan glucanohydro-lase | Rhizopus chinensis | β-glucan | 4800 | NA | NA |
| 1,6-β-dglucan glucanohydro-lase | 1,6-β-dglu-can glucanohy-drolase | β-glucan/gentiobiose/ pachyman | 405 | 50 | 4.2 |

In addition, hemicellulose degradation needs accessory enzymes such as xylan esterases, ferulic and p-coumaric esterases, α-1-arabinofuranosidases, and α-4-O-methyl glucuronosidases, acting synergistically to efficiently hydrolyze wood xylans and mannans. In the case of O-acetyl-4-O-methylglucuronxylan, which is one of the most common hemicelluloses, four different enzymes are required for degradation: endo-1-4-β-xylanase (endox-ylanase), acetyl esterase, α-glucuronidase and β-xylosidase. The degradation of O-acetylgalactoglucomannan starts with rupture of the polymer by endomannases. Acetylglucomannan esterases remove acetyl groups, and α-galactosidases eliminate galactose residues. Finally, β-mannosidase and β-glycosidase break down the endomannase-generated oligomeric β-1,4 bonds. Table shows the highest specific activity (μmol min− 1 mg− 1) reported for hemicellulases.

Table: Highest specificity activity of fungus hemicellulases.

| Enzyme | Organism | Substrates | Specific activity (μmol $min^{-1} mg^{-1}$) | Opt. T (°C) | Opt. pH |
|---|---|---|---|---|---|
| Feruloyl esterase | Aspergillus níge | Methyl sinapinate | 156 | 55 | 5 |
| Endo-1,4-β xylanase | Trichoderma longibrachiatum | 1,4-β-d-xylan | 6630 | 45 | 5 |
| β-1,4-Xylosidase | Aspergillus nidulans | ρ-nitrophenyl-β- dxylopy-ranoside | 107 | 50 | 5 |
| Exo-β-1,4-Mannosidase | Aspergillus níge | r β-d-Man-(1-4)-β-dG-lcNAc-(1-4)-β-dGlcNAc-Asn-Lys | 188 | 55 | 3.5 |
| Endo-β-1,4-mannanase | Sclerotium rolfsii | Galactoglucomannan/ mannans/ galactomannans/ glucomannans/ | 380 | 72-74 | 2.9/3.3 |
| Endo-α-1,5-arabinanase | Aspergillus Níger | 1,5-α-l-arabinan | 90 | 50 55 | 4.5- 5.0 |
| α-l Arabinofuranosidase | Aspergillus Níger | 1,5-α-l arabinofuranohexaose/ 1,5-α-larabinotriose/ 1,5-l-arabinan/α- larab-inofuranotriose | 397 | 50- 60 | 3.4- 4.5 |
| α-Glucuronidase | Phanerochaete Chrysosporium | 4-O-methyl glucuronosyl-xylobiose | 4.5 | 50 | 3.5 |
| α-Galactosidase | Mortierella vinacea | Melibiose | 2000 | 60 | 4 |
| Endo-galactanase | Aspergillus nige | NA | 6593 | 50-55 | 3.5 |
| β-glucosidase | Humicola insolvens | (2-hydroxymethylphenyl)-β-d-glucopyranoside | | | |

| Acetyl xylan | Schizophyllum | 4-methylumbelliferyl | 227 | 30 | 7.7 |
|---|---|---|---|---|---|
| esterase | commune | acetate/4-nitrophenyl acetate | | | |

## Generation of Lignocellulosic Residues

The increasing expansion of agro-industrial activity has led to the accumulation of a large quantity of lignocellulosic residues from wood (e.g. poplar trees), herbaceous (e.g. switchgrass), agricultural (e.g. corn stover, and wheat straw), forestry (e.g. sawdust, thinnings, and mill waste), municipal solid wastes (e.g. waste paper) and various industrial wastes all over the world. Table summarizes the worldwide generation of lignocellulosic residues.

## Bioconversion of Lignocellulose into Bioproducts

Bioconversion of lignocellulosic residues to useful, higher value products normally requires multi-step processes, which include:

1. Pretreatment (mechanical, chemical or biological);

2. Hydrolysis of polymers to produce readily metabolizable molecules (e. g. hexose or pentose sugars);

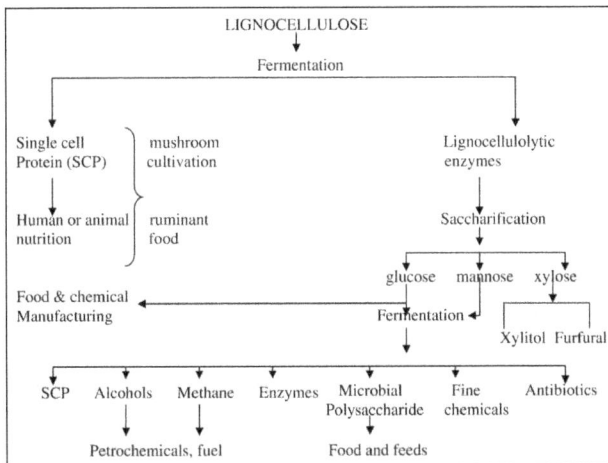

Generalized process stages in lignocellulosic waste bioconversion

3. Use of these molecules to support microbial growth or to produce chemical products; and

4. Separation and purification.

Several uses have been suggested for biodegrade lignocellulosic wastes; among them are used as raw material for the production of ethanol, for paper manufacturing, for compost making for cultivation of edible mushroom, and directly as animal feed.

Much research has been done in finding an alternative fuel using biological methods because of the positive environmental benefits of biofuels. Ethanol is either used as a chemical feedstock or as an additive to gasoline. Softwood, the dominant source of lignocellulose in the Northern hemisphere, has been the subject of interest as a raw material for fuel ethanol production in Sweden, Canada and Western USA. Ethanol fuel can reduce greenhouse gas emissions and improve air quality as well as offer strategic or economical advantages. These gasoline fuels contain up to 20% ethanol by volume.

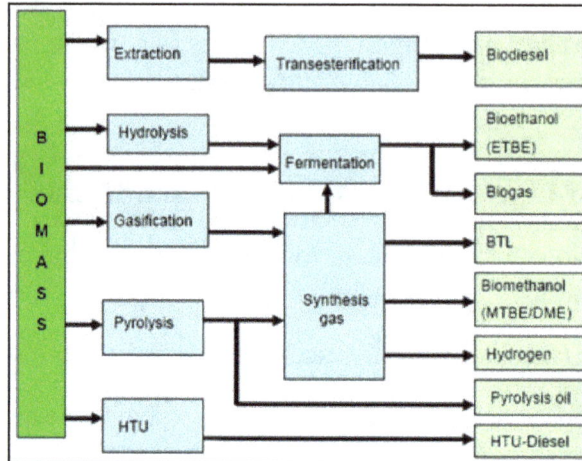

Biomass energy conversion overview

Over the past two decades, the cost of biological conversion of cellulosic biomass to ethanol has been reduced from around 1.22 USD l– 1 to the point where it is becoming economically viable. A number of high-value bioproducts such as organic acids, amino acids, vitamins and a number of bacterial and fungal polysaccharides such as xanthans are produced by fermenta- tion using glucose as the base substrate but theoretically these same products could be manufactured from "lignocellulosic residues". reported that based on the known metabolism of P. chrysosporium, several potential value-added products could be derived from lignin.

Table: Lignocellulosic residues generated from different agricultural sources.

| Lignocellulosic residues | Ton × $10^6$ /year |
|---|---|
| Sugar cane bagasse | 317-380 |
| Maize straw | 159-191 |
| Rice shell | 157-188 |
| Wheat straw | 154-185 |
| Soja straw | 54-65 |
| Yuca straw | 40-48 |

| Barley straw | 35-42 |
|---|---|
| Cotton fiber | 17-20 |
| Sorgoum straw | 15-18 |
| Banana waste | 13-15 |
| Mani shell | 9.2-11.1 |
| Sunflower straw | 7.5-9.0 |
| Bean straw | 4.9-5.9 |
| Rye straw | 4.3-5.2 |
| Pine waste | 3.8-4.6 |
| Coffee straw | 1.6-1.9 |
| Almond straw | 0.4-0.49 |
| Hazelnut husk | 0.2-0.24 |
| Sisal a henequen straw | 0.0077-0.093 |

Cultivation of edible mushrooms using lignocellulosic residues is a value addition process to convert these materials into human food. It is one of the most efficient biological ways by which these residues can be recycled. Mushrooms can be grown successfully on a wide variety of lignocellulosic residues such as cereal straws, banana leaves, sawdust, peanuts hulls, coffee pulp, soybean and cotton stalk, and almost any lignocellulosic substrate that has a substantial cellulose component. Rumen microorganisms convert cellulose and other plant carbohydrates in large amounts to acetic, propionic and butyric acids, which ruminant animals can use as energy and carbon sources); these microbes also have promise for commercial bioprocessing of lignocellulosic wastes anaerobically in liquid digesters. Cariello et al. reported that a mixed of endogenous microorganism (Bacillus subtillis, Pseudomonas fluorescens and Aspergillus fumigatus) accelerated the composting process in municipal solid wastes. Studies about a combination of an integrated system of composting, with bioinoculants (strains of Pleurotus sajor-caju, Trichoderma harzianum, Asper- gillus niger and Azotobacter chroococcum) and subsequent vermicomposting showed an accelerated composting process of wheat straw besides producing a nutrient-enriched compost. Figure shows several technologies for converting biomass that are commercial today. Biomass pyrolysis is a process by which a biomass feedstock is thermally degraded in the absence of air/oxygen. It is used for the production of solid (charcoal), liquid (tar and other organics) and gaseous products. These products are of interest as they are possible alternate sources of energy.

The study of pyrolysis is gaining increasing importance and has many advantages over other renewable and conventional energy sources. In the gasification process, the biomass is heated in an environment where the solid biomass breaks down to form a flammable gas. The biogas can be cleaned and filtered to remove problem chemical compounds. The gas can be used in more efficient power generation systems called combined cycles, which combine gas turbines and steam turbines to produce electricity. Anaerobic digestion is a commercially proven technology and is widely used for recycling and treating wet organic waste and waste waters. It is a type of fermentation that converts organic material into biogas, which mainly consists of methane (approximately 60%) and carbon dioxide (approximately 40%) and is comparable to landfill gas. Similar to gas produced via gasification, gas from anaerobic digestion can, after appropriate treatment, be burned directly for cooking or heating. It can be used in secondary conversion devices such as an internal combustion engine for producing electricity or shaft work.

# Industrial Enzymes

Enzymes which are used for commercial purposes in various industries are known as industrial enzymes. Pharmaceuticals, biofuels, consumer products and chemical production are a few industries which make use of these enzymes. The chapter closely examines these industrial enzymes to provide an extensive understanding of the subject.

Industrial enzymes are used in a variety of applications - e.g. detergents, animal feeds, food processing, tanning, starch processing and textiles - with the highest volume applications in the detergent industry. The dominant producers in industrial enzymes are Novozymes and Genencor, which hold over 60% of the worldwide market. Both are seeking to diversify out of industrial enzymes, using their knowledge of protein expression and function to enter the more lucrative biotherapeutic sector.

## Types of Industrial Enzymes

## Microbial Cellulases

Microbial cellulases, a major group of industrial enzymes, have a wide range of applications as far as biotechnological, environmental issues, industrial products, and processes are concerned. Thermophilic bacteria act as a good source for industrial cellulase because of its compatibility during industrial harsh processes, for example, deinking of paper, softening of fabric material, pulp and paper, biostoning and biopolishing of fabric materials, juice, and animal feed, etc. Genome mediated application of thermophilic bacteria in various fields has become a rising fantasy among biotechnologists in the current scenario. The genetic manipulation of thermophilic bacteria may lead to enhanced cellulase production through recombinant technology, an emerging technology towards cellulase gene isolation of thermophilic bacteria and their expression in suitable hosts for enhanced cellulase production. Cellulase gene isolation from various thermophilic bacteria could be done through various approaches, such as a classical approach, whole genome isolation, whole metagenome isolation, and bioinformatics.

## Pectinolytic Enzymes

Pectinolytic enzymes are among the most important industrial enzymes with (potential) applications in cotton scouring, flax processing (retting) and fruit juice clarification. Like cellulases, pectinases are widespread in nature. Pectinases are produced in different micro-organisms as well as in plants. Pectinases are a group of enzymes (lyases as well as hydrolases) that catalyse the degradation of pectin, and are classified into three groups: pectin esterases, polygalacturonases and pectin lyases. Pectin esterase de-esterifies polymethyl-galacturonate forming polygalacturonate. Polygalacturonase hydrolyses α-(1–4) linkages in polygalacturonate, and can be divided in two groups according to the following criterion: whether cleavage occurs randomly within the polymer (endopolygalacturonase,) or end-wise from the non-reducing end of the polymer (exopolygalacturonase,). Pectin lysases remove double bonds in polygalacturonate via β-elimination, thereby cleaving the polymer. Three types of pectin lyases exist, namely: endopolygalacturonate lyases which randomly cleave polygalacturonate; exopolygalacturonate lyases which cleave polygalacturonate chains at the ends yielding unsaturated galacturonic acid; and endopolymethylgalacturonate lyases which randomly cleave pectin.

With increasing knowledge and understanding of pectinolytic enzymes and of their different pectin degradation mechanisms, alkaline pectate lyases seem to have great potential in cotton scouring and enzymatic flax retting. Actual commercialisation of enzymatic cotton scouring is still limited because of the relatively long incubation time, residual pectin content and the processing temperature required (~60 °C). The enzymatically scoured fabrics have a softer hand than those scoured conventionally, however the degree of whiteness is often somewhat lower and the process is not yet suitable for adequate removal of seed coat fragments and motes.

The purpose of scouring is to clean the surface of cotton fibres and to remove hydrophobic and non-cellulose parts of the cotton, rendering the surface of the fibres sufficiently hydrophilic. These hydrophobic components and non-cellulosic materials present in the cuticle and the primary wall of cotton have a role in the protection of the plant from the environment and pathogens as well as a structural role. In the conventional process, hydrophobic materials are removed in a non-specific manner and this involves the use of relatively harsh chemical solutions and high temperature.

## α-Amylase

Starch hydrolyzing enzymes, such as amylases, are among the most important industrial enzymes, and account for over 25% of industrial enzymes. α-Amylases are a varied family of enzymes that randomly cleave the α-1,4 linkages between adjacent glucose units in the linear amylose and amylopectin chain of starch enzymes. These enzymes have been the focus of many studies to improve their properties, in order to make them more amenable for industrial use.

Due to their intrinsic modularity, a wide range of chimeragenesis studies have been carried out with α-amylases. The starch-binding domain was exchanged between different species, improving the starch digesting functionalities.

## Enzymes as a Unit Operation

## Immobilization

Despite their excellent catalytic capabilities, enzymes and their properties must be improved prior to industrial implementation in many cases. Some aspects of enzymes that must be improved prior to implementation are stability, activity, inhibition by reaction products, and selectivity towards non-natural substrates. This may be accomplished through immobilization of enzymes on a solid material, such as a porous support. Immobilization of enzymes greatly simplifies the recovery process, enhances process control, and reduces operational costs. Many immobilization techniques exist, such as adsorption, covalent binding, affinity, and entrapment. Ideal immobilization processes should not use highly toxic reagents in the immobilization technique to ensure stability of the enzymes. After immobilization is complete, the enzymes are introduced into a reaction vessel for biocatalysis.

## Adsorption

Enzyme adsorption onto carriers functions based on chemical and physical phenomena such as van der Waals forces, ionic interactions, and hydrogen bonding. These forces are weak, and as a result, do not affect the structure of the enzyme. A wide variety of enzyme carriers may be used. Selection of a carrier is dependent upon the surface area, particle size, pore structure, and type of functional group.

## Covalent Binding

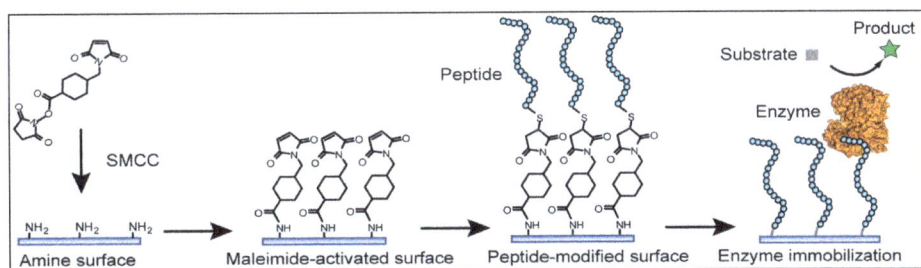

Example of Enzyme Immobilization through Covalent Binding

Many binding chemistries may be used to adhere an enzyme to a surface to varying degrees of success. The most successful covalent binding techniques include binding via glutaraldehyde to amino groups and N-hydroxysuccinide esters. These immobilization techniques occur at ambient temperatures in mild conditions, which have limited potential to modify the structure and function of the enzyme.

## Affinity

Immobilization using affinity relies on the specificity of an enzyme to couple an affinity ligand to an enzyme to form a covalently bound enzyme-ligand complex. The complex is introduced into a support matrix for which the ligand has high binding affinity, and the enzyme is immobilized through ligand-support interactions.

## Entrapment

Immobilization using entrapment relies on trapping enzymes within gels or fibers, using non-covalent interactions. Characteristics that define a successful entrapping material include high surface area, uniform pore distribution, tunable pore size, and high adsorption capacity.

## Recovery

Enzymes typically constitute a significant operational cost for industrial processes, and in many cases, must be recovered and reused to ensure economic feasibility of a process. Although some biocatalytic processes operate using organic solvents, the majority of processes occur in aqueous environments, improving the ease of separation. Most biocatalytic processes occur in batch, differentiating them from conventional chemical processes. As a result, typical bioprocesses employ a separation technique after bioconversion. In this case, product accumulation may cause inhibition of enzyme activity. Ongoing research is performed to develop *in situ* separation techniques, where product is removed from the batch during the conversion process. Enzyme separation may be accomplished through solid-liquid extraction techniques such as centrifugation or filtration, and the product-containing solution is fed downstream for product separation.

| Enzymes as a Unit Operation | | |
|---|---|---|
| **Enzyme** | **Industry** | **Application** |
| Palatase | Food | Enhance cheese flavor |
| Lipozyme TL IM | Food | Interesterification of vegetable oil |
| Lipase AK Amano | Pharmaceutical | Synthesis of chiral compounds |
| Lipopan F | Food | Emulsifier |
| Cellulase | Biofuel | Class of enzymes that degrade cellulose to glucose monomers |
| Amylase | Food/biofuel | Class of enzymes that degrade starch to glucose monomers |
| Xylose isomerase | Food | High fructose corn syrup production |
| Resinase | Paper | Pitch control in paper processing |
| Penicillin amidase | Pharmaceutical | Synthetic antibiotic production |
| Amidase | Chemical | Class of enzymes used for non-proteinogenic entiomericically pure amino acid production |

## Enzymes as a Desired Product

To industrialize an enzyme, the following upstream and downstream enzyme production processes are considered:

## Upstream

Upstream processes are those that contribute to the generation of the enzyme.

1. Selection of a suitable enzyme

An enzyme must be selected based upon the desired reaction. The selected enzyme defines the required operational properties, such as pH, temperature, activity, and substrate affinity.

2. Identification and selection of a suitable source for the selected enzyme

The choice of a source of enzymes is an important step in the production of enzymes. It is common to examine the role of enzymes in nature, and how they relate to the desired industrial process. Enzymes are most commonly sourced through bacteria, fungi, and yeast. Once the source of the enzyme is selected, genetic modifications may be performed to increase the expression of the gene responsible for producing the enzyme.

3. Process development

Process development is typically performed after genetic modification of the source organism, and involves the modification of the culture medium and growth conditions. In many cases, process development aims to reduce mRNA hydrolysis and proteolysis.

4. Large scale production

Scaling up of enzyme production requires optimization of the fermentation process. Most enzymes are produced under aerobic conditions, and as a result, require constant oxygen input, impacting fermenter design. Due to variations in the distribution of dissolved oxygen, as well as temperature, pH, and nutrients, the transport phenomena associated with these parameters must be considered. The highest possible productivity of the fermenter is achieved at maximum transport capacity of the fermenter.

## Downstream

Downstream processes are those that contribute to separation or purification of enzymes.

1. Removal of insoluble materials and recovery of enzymes from the source

The procedures for enzyme recovery depend on the source organism, and whether enzymes are intracellular or extracellular. Typically, intracellular enzymes require cell lysis and separation of complex biochemical mixtures. Extracellular enzymes are released into the culture medium, and are much simpler to separate. Enzymes must maintain their native conformation to ensure their catalytic capability. Since enzymes are very sensitive to pH, temperature, and ionic strength of the medium, mild isolation conditions must be used.

2. Concentration and primary purification of enzymes

Depending on the intended use of the enzyme, different levels purity are required. For example, enzymes used for diagnostic purposes must be separated to a higher purity than bulk industrial enzymes to prevent catalytic activity that provides erroneous results. Enzymes used for therapeutic purposes typically require the most rigorous separation. Most commonly, a combination of chromatography steps is employed for separation.

The purified enzymes are either sold in pure form and sold to other industries, or added to consumer goods.

| Enzymes as a Desired Product | | |
|---|---|---|
| Enzyme | Industry | Application |
| Novozym-435 | Consumer Goods | Isopropyl myristate production (Cosmetic) |
| Bromelain | Food | Meat tenderizer |
| Noopazyme | Food | Improve noodle quality |
| Asparaginase | Pharmaceutical | Lymphatic cancer therapeutic |
| Ficin | Pharmaceutical | Digestive aid |
| Urokinase | Pharmaceutical | Anticoagulant |
| β-Lactamase | Pharmaceutical | Penicillin allergy treatment |
| Subtilisin | Consumer Goods | Laundry detergent |

## Use of Industrial Enzymes

Enzymes are used in industrial processes, such as baking, brewing, detergents, fermented products, pharmaceuticals, textiles, leather processing.

## Pharmaceutical and Analytical Industry

Enzymes have many significant and vital roles in the pharmaceutical and diagnostic industries. These are extensively used as therapeutic drugs in health issues associated with enzymatic deficiency and digestive disorders, and in diagnostic procedures such as ELISA and diabetes testing kits.

Enzyme applications in medicine are as extensive as in industry and are growing rapidly. At present, most prominent medical uses of microbial enzymes are removal of dead skin, and burns by proteolytic enzymes, and clot busting by fibrinolytic enzymes. Nattokinase, a potent fibrinolytic enzyme, is a promising agent for thrombosis therapy.

Acid protease, dextranase and rhodanase may be used to treat alimentary dyspepsia, tooth decay and cyanide poisoning, respectively. Lipases are the most frequently used enzymes in the organic synthesis and are used in the synthesis of optically active alcohols, acids, esters, and lactones. Microbial lipases and polyphenol oxidases are involved in the synthesis of (2R,3S)-3-(4-methoxyphenyl) methyl glycidate (an intermediate for diltiazem) and 3, 4-dihydroxylphenyl alanine (DOPA, for treatment of Parkinson's disease), respectively. Tyrosinase, an important oxidase enzyme, is involved in melanogenesis and in the production of l-Dihydroxy phenylalanine (L-DOPA). L-DOPA is used as a precursor for the production of dopamine which is a potent drug for the treatment of Parkinson's disease and to control the myocardium neurogenic injury. Chitosanase catalyze hydrolysis of chitosan to biologically active chitosan oligosaccharides (COS), which is used as antimicrobial, antioxidant, lowering of blood cholesterol and high blood pressure, controlling arthritis, protective effects against infections and improving antitumor properties. Applications of microbial enzymes for different health problems are illustrated in table.

Table: Some therapeutic applications of microbial enzymes.

| Treatment | Enzymes | Microorganisms |
|---|---|---|
| Antitumor | L-Asparaginase, L-glutaminase, L-tyrosinase, galactosidase | Escherichia coli, Pseudomonas acidovorans, Beauveria bassiana, Acinetobacter |
| Antiinflammatory | Superoxide dismutase, Serrapeptase | Lactobacillus plantarum, Nocardia sp., Mycobacteriumsp., Corynebacterium Glutamicum, |
| Anticoagulants | Streptokinase, urokinase | Streptococci sp., Bacillus subtilis |
| Antibiotic synthesis | Penicillin oxidase, rifamycin B oxidase | Penicillium sp. |
| Antioxidants | Superoxide dismutases, glutathione peroxidases, catalase | Lactobacillus plantarum, Corynebacterium glutamicum |
| Skin ulcers | Collagenase | Clostridium perfringens |
| Detoxification | Laccase, rhodanese | Pseudomonas aeruginosa |
| Antibiotic resistance | β-Lactamase | Klebsiella pneumonia, Citrobacter freundii, Serratia marcescens |
| Antiviral | Ribonuclease, Serrapeptase | Saccharomyces cerevisiae |
| Gout | Uricase | Aspergillus flavus |
| Digestive disorders | α-Amylase, lipase | Bacillus spp., Candida lipolytica, A. oryzae |
| Cyanide poisoning | Rhodanase | Sulfobacillus sibiricus |

The extensive utilization of enzymes for scientific and analytical purposes is used to estimate the concentration of substrates and to determine the catalytic activity of enzymes present in biological samples. Advances in the enzyme technology have replaced or minimized the use to harmful radioactive elements in different immunoassays, which are used for the determination of a variety of proteins and hormones.

Furthermore, enzymes are used in clinical diagnostic for the quantitative determination of diabetes and other health disorders, for example, glucose oxidase for glucose;

urease and glutamate dehydrogenase for urea; lipase, carboxyl esterase, and glycerol kinase for triglycerides; urate oxidase for uric acid; creatinase and sarcosine oxidases for creatinine. Cholesterol oxidase has also been reported for useful biotechnological applications in the detection and conversion of cholesterol. Putrescine oxidase is used to detect biogenic amines, such as putrescine, a marker for food spoilage.

Enzymes are indispensable in nucleic acid manipulation for research and development in the field of genetic engineering, such as restriction endonucleases are used for site specific cleavage of DNA for molecular cloning and DNA polymerases for the DNA amplification by polymerase chain reaction (PCR).

## Food Industry

These biomolecules are efficiently involved in improving food production and components, such as flavor, aroma, color, texture, appearance and nutritive value. The profound understanding of the role of enzymes in the food manufacturing and ingredients industry have improved the basic processes to provide better markets with safer and higher quality products. Furthermore, the enzymes gained interest in new areas such as fat modification and sweetener technology. In beverage and food industry, enzymes are added to control the brewing process and produce consistent, high-quality beer; to enhance the functional and nutritional properties of animal and vegetables proteins by the enzymatic hydrolysis of proteins, for higher juice yield with improved color and aroma.

The application of enzymes in food industry is segmented into different sectors, such as baking, dairy, juice production and brewing. Worldwide, microbial enzymes are efficiently utilized in bakery—the principal application market in food industry—to improve dough stability, crumb softness and structure, and shelf life of products. Increased uses of microbial enzymes in cheese processing are largely responsible for the use of enzymes in dairy industry, which is the next largest application industry followed by the beverages industry.

Baking industry- Baking enzymes are used for providing flour enhancement, dough stability, improving texture, volume and color, prolonging crumb softness, uniform crumb structure and prolonging freshness of bread. To meet rising demand for quality, enzymes are seen as natural solutions in today's baking market.

Bread making is one of the most common food processing techniques globally. The use of enzymes in bread manufacturing shows their value in quality control and efficiency of production. Amylase, alone or in combination with other enzymes, is added to the bread flour for retaining the moisture more efficiently to increase softness, freshness and shelf life. Additionally, lipase and xylanase are used for dough stability and conditioning while glucose oxidase and lipoxygenase added to improve dough strengthening and whiteness. Transglutaminase is used in baking industry to enhance the quality of flour, the amount and texture of bread and the texture of cooked pasta. Lipases are also used to improve the flavor content of bakery products by liberating short-chain fatty acids through esterification and to prolong the shelf life of the bakery products.

Dairy industry Dairy enzymes, an important segment of food enzyme industry, are used for the development and enhancing organoleptic characteristics (aroma, flavor and color) and higher yield of milk products. The use of enzymes (proteases, lipases, esterases, lactase, aminopeptidase, lysozyme, lactoperoxidase, transglutaminase, catalase, etc.) in dairy market is well recognized and varies from coagulant to bio-protective enzyme to enhance the shelf life and safety of dairy products. Dairy enzymes are used for the production of cheese, yogurt and other milk products.

Rennet, a combination of chymosin and pepsin, is used for coagulation of milk into solid curds for cheese production and liquid whey. Currently, approximately 33 % of global demand of cheese produced using microbial rennet. Other proteases find applications for accelerated cheese processing and in reduction of allergenic properties of milk products. Currently, lipases are involved in flavor improvement, faster cheese preparation, production of customized milk products, and lipolysis of milk fat. Transglutaminase catalyzes polymerization of milk proteins and improves the functional properties of dairy products.

Lactose intolerance is the lack of ability of human being to digest lactose due to deficiency of lactase enzyme. Lactase catalyzes hydrolysis of lactose to glucose and galactose, and therefore, is used as a digestive aid and to enhance the solubility and sweetness in milk products. It is required to minimize or removal of lactose content of milk products for lactose-intolerant people to prevent severe tissue dehydration, diarrhea, and sometimes fatal consequences.

Beverages industry The beverage industry is divided into two major groups and eight sub-groups. The nonalcoholic group contains soft drink and syrup, packaged water, fruit juices along with tea and coffee industry. Alcoholic group comprised distilled spirits, wine and beer. Industrial enzymes are used in breweries as processing aids and to produce consistent and high-quality products. In the brewing industries, microbial enzymes are used to digest cell wall during extraction of plant material to provide improved yield, color, and aroma and clearer products.

The enzyme applications are an integrating ingredient of the current fruit and vegetable juice industry. Enzymes are used in fruit and vegetable juice industry as processing aids to increase the efficiency of operation, for instance, peeling, juicing, clarification, extraction and improve the product quality. Application of cellulases, amylases, and pectinases during fruit juice processing for maceration, liquefaction, and clarification, improve yield and cost effectiveness. The quality and stability of juices manufactured are enhanced by the addition of enzymes. Enzymes digest pectin, starch, proteins and cellulose of fruits and vegetables and facilitate improved yields, shortening of processing time and enhancing sensory characteristics. Amylases are used for clarification of juices to maximize the production of clear or cloudy juice. Cellulases and pectinases are used to improve extraction, yield, cloud stability and texture in juices. Naringinase and limoninase, debittering enzymes, hydrolyze bitter components and improves the quality attributes of citrus juices. Pectin, a structural heteropolysaccharide, present in nearly all fruits is required to be maintained to regulate cloudiness of juices by polygalacturonase, pectin esterases pectin lyase and various arabanases.

Microbial amylases may be utilized in the distilled alcoholic beverages to hydrolyze starch to sugars prior to fermentation and to minimize or remove turbidities due to starch. The application of enzymes to hydrolyze unmalted barley and other starchy adjuncts facilitate in cost reduction of beer brewing. In brewing, development of chill-hazes in beer may be control by the addition of proteases.

1. Feed industry

Demand of milk and meat consumption, growth of feed enzymes occurred steadily. The use of enzymes in animal diets initiated in the 1980s and exploded in the 1990s. Feed enzymes are gaining importance as they can increase the digestibility of nutrients and higher feed utilization by animals. The global market for feed enzymes was estimated $899.19 million in 2014 and expected to reach nearly $1.3 billion by 2020, at a CAGR of 7.

Feed enzymes may be used in animal diet formulation. For instance, these are added to degrade specific feed components which are otherwise harmful or no nutritional value. In addition, the protein dietary value of feeds available for poultry may also be enhanced by the application of feed enzymes. Feed enzymes mainly used for poultry are phytases, proteases, α-galactosidases, glucanases, xylanases, α-amylases, and polygalacturonases. The phytase, largest enzyme segment in the feed industry, is used to utilize natural phosphorous bound in phytic acid in cereal-based feed. Monogastric animals are unable to digest plant based feeds containing high amount of cellulose and hemicelluloses. Xylanase and β-glucanase are added to their feeds as these enzymes fully degrade and digest high amount of starch. Proteases are also used in animal feeds to overcome anti-nutritional factors by degrading proteins into their constituent amino acids. Apart from improving the nutritional value of feed for better feed conversion by the animals, these feed enzymes are gaining importance for their role in feed cost reduction and meat quality improvement.

2. Polymer industry

To meet the increased consumption of polymers and the growing concern for human health and environmental safety has led to the utilization of microbial enzymes for synthesis of biodegradable polymer. In vitro enzyme catalyzed synthesis of polymer is an environmental safe process having several advantages over conventional chemical methods. Biopolymers are environmentally friendly materials as these are synthesized from renewable carbon sources via biological processes, degrade biologically after use and return to the natural environment as renewable resources, such as $CO_2$

and biomass. Biopolymers, such as polyesters, polycarbonates and polyphosphates are used in various biomedical applications, e.g., orthopedic devices, tissue engineering, adhesion barriers, control drug delivery, etc.

## 3. Paper and Pulp industry

With increasing awareness of sustainability issues, uses of microbial enzymes in paper and pulp industry have grown steadily to reduce adverse effect on ecosystem. The utilization of enzymes reduce processing time, energy consumption and amount of chemicals in processing. Enzymes are also used to enhance deinking, and bleach in paper and pulp industry and waste treatment by increasing biological oxygen demand (BOD) and chemical oxygen demand (COD). Xylanases and ligninases are used in paper and pulp industries to augment the value of the pulp by removing lignin and hemicelluloses. In these industries, amylases uses include starch coating, deinking, improving paper cleanliness and drainage improvement. Lipases are employed for deinking and enhancing pitch control while cellulases are used for deinking, improving softness and drainage improvement. Cellulase has also been used for the development of the bioprocess for recycling of used printed papers. The application of laccase is an alternative to usage and requirement of large amount of chlorine in chemical pulping process; subsequently, reduce the waste quantity that causes ozone depletion and acidification. Moreover, mannases are used for degrading glucomannan to improve brightness in paper industry.

## 4. Leather industry

The leather industry is more customary, and therefore, discharges and waste disposed from different stages of leather processing are causing severe health hazards and environmental problems. The biodegradable enzymes are efficient alternative to improve the quality of leather and help to shrink waste. The initial attempt for application of enzyme in leather industry was made for dehairing process, the largest process in leather preparation which require bulk amount of enzymes like proteases, lipases and amylases. Enzymatic dehairing applications are attractive because it can preserve the hair and contribute to fall in the organic load released into the effluent. Enzymatic dehairing processes minimize or eliminate the dependence on harmful chemicals, such as sulfide, lime and amines.

Enzymes are required for facilitating procedure and enhancing leather quality during different stages in leather processing, such as, curing, soaking, liming, dehairing, bating, picking, degreasing and tanning. The enzymes used in leather industries are alkaline proteases, neutral proteases, and lipases. Alkaline proteases are used to remove non fibrillar proteins during soaking, in bating to make leather soft, supple and pliable. Neutral and alkaline proteases, both are used in dehairing to reduce water wastage. In addition to this, lipases are used during degreasing to remove fats. The advantages of using enzymes instead of chemicals in liming are stainless pelt, reduced odor, low BOD and COD in effluents, and improved hair recovery.

## 5. Textile industry

The textile industry is responsible for vast generation of waste from desizing of fabrics, bleaching chemicals and dye is one of the largest contributors to environmental pollution. In such industries, enzymes are used to allow the development of environmentally friendly technologies in fiber processing and strategies to improve the final product quality. The main classes of enzymes involved in cotton pre-treatment and finishing processes are hydrolase and oxidoreductase. The group of hydrolase includes amylase, cellulase, cutinase, protease, pectinase and lipase/esterase, which are involved in the biopolishing and bioscouring of fabric, anti-felting of wool, cotton softening, denim finishing, desizing, wool finishing, modification of synthetic fibers, etc.. Oxidoreductase, other group of enzyme, includes catalase, laccase, peroxidase, and ligninase, which are involved in bio-bleaching, bleach termination, dye decolorization, fabric, wool finishing, etc.. A brief detail of applications of enzymes in textiles industries are shown in table.

Table: Uses of enzymes in textile industry.

| Enzyme | Use | Microorganisms |
| --- | --- | --- |
| Amylase | Desizing | Bacillus sp., B. licheniformis |
| Cellulose | Cotton softening, denim finishing | Aspergillus niger, Penicillium funiculosum |
| Catalase | Bleach termination | Aspergillus sp. |
| Laccase | Non-chlorine Bleaching, fabric dyeing | Bacillus subtilis |
| Pectate lyase | Bioscouring | Bacillus sp., Pseudomonas sp. |
| Amylase | Desizing | Bacillus sp., B. licheniformis |
| Cellulose | Cotton softening, denim finishing | Aspergillus niger, Penicillium funiculosum |
| Protease | Removal of wool fiber scales, degumming of silk | Aspergillus niger, B. subtilis |
| Lipase | Removal of size lubricants, denim finishing, | Candida Antarctica |
| Ligninase | Wool finishing | Trametes versicolor, Phlebia radiata |
| Collagenase | Wool finishing | Clostridium histolyticum |
| Cutinase | Cotton scouring, synthetic fiber modification | Pseudomonas mendocina, Fusarium solani pisi, Thermomonospora fusca |

## Enzymes in Cosmetics

The applications of enzymes in cosmetics have been continuously increased. Enzymes are used as free radical scavengers in sunscreen cream, toothpaste, mouthwashes, hair waving and dyeing. The superoxide dismutase is used to arrest free radicals and to control damage to skin caused by air and water pollutions, microbes and other harmful factors. SOD and peroxidases are used in combination in sunscreen cream as free radical scavengers to reduce erythema. Proteases are used in skin creams to clean and smoothen the skin through peeling off dead or damaged skin.

Other widely used enzymes in toothpaste and mouthwash are endoglycosidase and papain, which are used to whiten teeth, to remove plaque and to remove odor-causing deposits on teeth and gum tissue. Laccase, oxidases, peroxidases, and polyphenol oxidases are used in hair dyeing, lipase, catalase, papain, bromelain and subtilisin in skin care; and protein disulfide isomerase, glutathione sulfhydryl oxidase and transglutaminase in hair waving. Additionally, enzymes are also used in contact lens cleaners to remove protein films.

## Enzymes in Detergents

Enzymes have contributed significantly to the growth and development of industrial detergents, which is a prime application area for enzymes today. Detergents are used in miscellaneous applications as dishwashing, laundering, domestic, industrial and institutional cleaning. The enzymes in detergent products are used to remove protein, starch, oil and fats based stains and to increase the effectiveness of detergents. The enzymes in laundry detergents are weight efficient, cleave off damaged cotton fibers, improve whiteness, color and fabric care. Enzymes mainly used in detergent products are of hydrolase group and currently, most commonly used enzymes are amylase and protease. Sometimes a combination of enzymes, including proteases, amylases, pectinases, cellulases and lipases used to increase efficiency on stain cleaning and fabric care.

Amylases and lipases are effective on removing starchy food deposits and stains resulting from fatty products, respectively. Cutinase, a hydrolytic enzyme, is used as a lipolytic enzyme in dishwashing and laundry detergents. Protease digests on organic stains, such as grass, blood, egg and human sweat, whereas cellulases are used to brighten colors, soften fabrics and to eliminate small fibers from the fabric without damaging the major fibers of the fabric. Protease and amylase are used particularly in dishwasher detergents to remove protein and carbohydrate containing food particles. The application of enzymes in detergent products is advantageous as these products contain less bleaching agents, phosphates, and consequently have beneficial effects on public and environmental health.

## Organic Synthesis Industry

Enzyme based processes for production of fine chemicals are rapidly gaining practical significance owing to more economical high purity products in an eco-environmentally acceptable manner. Enzymes are preferred in industrial chemical synthesis over conventional methods for their high selectivity, i.e., chiral, positional and functional group specific. Such high selectivity is extremely advantageous in chemical synthesis as it may offer several benefits such as minimal or no by-product formation, easier separation, and less environmental problems. Besides, high catalytic efficiency and mild operational conditions are advantages of enzyme mediated commercial applications. Catalytic potential of microorganisms have been employed for hundreds of years in the production of alcohol, and cheese for industrial synthetic chemistry. Among the enzymes in organic synthesis, lipases are the most frequently used, particularly, in the formation of a wide range of optically active alcohols, acids, esters, and lactones. Lipases are used for the production of (S, R)-2, 3-p-ethoxyphenylglycyclic acid, an intermediate for diltiazem. Oxidoreductases, such as polyphenol oxidase is involved in the synthesis of 3,4-dihydroxylphenyl alanine (DOPA), a chemical used in the treatment of Parkinson's disease. Oligosaccharides and polysaccharides, play vital roles in cellular recognition and communication processes, are synthesized industrially using high regio- and stereoselectivity of glycosyltransferases. Lyases are involved in organic synthesis of cyanohydrins from ketones, acrylamide from acrylonitrile, malic acid from fumaric acid.

## Waste Treatment

The use of enzyme for waste management is extensive and a number of enzymes are involved in the degradation of toxic pollutants. The industrial effluents as well as domestic waste contain many chemical commodities, which are hazardous or toxic to the living being and ecosystem. Microbial enzyme(s), alone or in combinations, are used for the treatment of industrial effluents containing phenols, aromatic amines, nitriles, etc., by degradation or bioconversion of toxic chemical compound(s) to innocuous products. A number of enzymes employed for waste treatment are amidases, amylases, amyloglucosidases, cellulases, glucoamylases, lipases, nitrile hydratases, pectinases and proteases. The detoxification of toxic organic compounds through oxidative coupling is mediated with oxidoreductases. These enzymes, like laccase, manganese peroxidase, lignin peroxidase and tyrosinase catalyze the removal of chlorinated phenolic compounds from industrial effluents. The microbial enzymes are also utilized to recycle the waste for reuse, e.g., to recover additional oil from oil seeds, to convert starch to sugar, to convert whey to various useful products. Microbial oxygenases, such as monooxygenases and dioxygenases have a broad substrate range, and are active against a wide range of compounds, including the chlorinated aliphatics. These are used in the degradation of halogenated organic compounds containing pollutants, like herbicides, insecticides, fungicides, hydraulic and heat transfer fluids, plasticizers, and intermediates for chemical synthesis.

## Immobilized Enzyme

Immobilized enzymes are enzymes that have been attached, either through adsorption or crosslinking, to a solid support for experimentation or industrial purposes; for example, generating large quantities of products through biosynthesis. The inert, water-insoluble calcium alginate, as well as water-soluble matrices, can serve as the immobilization substrate.

# Enzyme Immobilization

Immobilization of enzymes (or cells) refers to the technique of confining/anchoring the enzymes (or cells) in or on an inert support for their stability and functional reuse. By employing this technique, enzymes are made more efficient and cost-effective for their industrial use.

Salient Features of enzyme immobilization:

1. The enzyme phase is called as carrier phase which is water insoluble but hydrophilic porous polymeric matrix, e.g. agarose, cellulose, etc.

2. The enzyme phase may be in the form of fine particulate, membranous, or microcapsule.

3. The enzyme in turn may be bound to another enzyme via cross linking.

4. A special module is produced employing immobilization techniques through which fluid can pass easily, transforming substrate into product and at the same time facilitating the easy removal of catalyst from the product as it leaves the reactor.

5. The support or carrier utilized in immobilization technique is not stable at particular pH, ionic strength, or solvent conditions. Hence, may be disrupted or dissolved releasing the enzyme component after the reaction.

Advantages of enzyme immobilization:

- Multiple or repetitive use of a single batch of enzymes.

- Immobilized enzymes are usually more stable.

- Ability to stop the reaction rapidly by removing the enzyme from the reaction solution.

- Product is not contaminated with the enzyme.

- Easy separation of the enzyme from the product.

- Allows development of a multienzyme reaction system.

- Reduces effluent disposal problems.

Disadvantages of enzyme immobilization:

- It gives rise to an additional bearing on cost.

- It invariably affects the stability and activity of enzymes.

- The technique may not prove to be of any advantage when one of the substrate is found to be insoluble.

- Certain immobilization protocols offer serious problems with respect to the diffusion of the substrate to have an access to the enzyme.

## Technique of Enzyme Immobilization

1. Carrier binding

   - Physical adsorption

   - Covalent bonding

   - Ionic bonding

2. Cross linking

3. Entrapment

   - Occlusion within a cross linked gel

   - Microencapsulation

## Physical Adsorption

This method is based on the physical adsorption of enzyme protein on the surface of water-insoluble carriers. Examples of suitable adsorbents are ion-exchange matrices, porous carbon, clay, hydrous metal oxides, glasses and polymeric aromatic resins.

The bond between the enzyme and carrier molecule may be ionic, covalent, hydrogen, coordinated covalent or even combination of any of these.

Immobilization can be brought about by coupling an enzyme either to external or internal surface of the carrier.

The external surface binding method is advantageous as it does not involve conditions like pore diffusion. The disadvantages, however, include exposure of enzymes to microbial attack, physical abrasion of enzyme due to turbulence associated with the bulk solution.

The major disadvantage of the internal immobilization method is the pore diffusion.

Advantages of adsorption:

- Little or no confirmation change of the enzyme.

- Simple and cheap.

- No reagents are required.

- Wide applicability and capable of high enzyme loading.

Disadvantages of adsorption:

- Desorption of the enzyme protein resulting from changes in temperature, pH, and ionic strength.

- Slow method.

## Methods of Immobilization by Adsorption

The absorptive immobilization of enzymes can be done by following methods:

1.  Static Process: This is most efficient technique but requires maximum time. In this technique, enzyme is immobilized by allowing it to be in contact with the carrier without agitation.

2.  Dynamic Process: This process typically involves the admixing of enzyme with the carrier under constant agitation using mechanical shaker.

3.  Reactor loading: This process is employed for the commercial production of immobilized enzymes. The carrier is placed into the reactor and enzyme solution is transferred to the reactor with agitation of the whole content in the reactor.

4.  Electro-Deposition: In this technique, carrier is placed in the vicinity of one of the electrode in an enzyme bath and electric current is applied leading to migration of enzyme towards the carrier. This results in deposition of enzyme on the surface of the carrier.

### Covalent Bonding

Covalent binding is the most widely used method for immobilizing enzymes. The covalent bond between enzyme and a support matrix forms a stable complex. The functional group present on enzyme, through which a covalent bond with support could be established, should be non essential for enzymatic activity.

The most common technique is to activate a cellulose-based support with cyanogen bromide, which is then mixed with the enzyme.

The protein functional groups which could be utilized in covalent coupling include:

- Amino group

- Carboxylic group

- Phenol ring

- Indole group

- Imidazole group

On the other hand examples of the polymeric supports include:

- Amino and related groups of polysaccharides and silica gel etc.

- Carboxylic acid and related groups of polyglutamic acid, carboxy methyl cellulose.

- Aldehyde and acetal groups of polymers.

- Amide gr. of polypeptide.

The polymers may be engaged in direct coupling as well as could be modified by other coupling groups or activating groups. The most commonly used polymers are polysaccharides, polyvinyl alcohol, silica and porous glasses.

Advantages of covalent coupling:

- The strength of binding is very strong, so, leakage of enzyme from the support is absent or very little.

- This is a simple, mild and often successful method of wide applicability

Disadvantages of covalent coupling:

- Enzymes are chemically modified and so many are denatured during immobilization.

- Only small amounts of enzymes may be immobilized (about 0.02 grams per gram of matrix).

## Cross Linking

This method is based on the formation of covalent bonds between the enzyme molecules, by means of multifunctional reagents, leading to three dimensional cross linked aggregates.

The most common reagent used for cross-linking is glutaraldehyde.

Advantages of cross linking:

- Very little desorption(enzyme strongly bound)

- Best used in conjunction with other methods.

Disadvantages of cross linking:

- Cross linking may cause significant changes in the active site.

## Entrapment

In entrapment, the enzymes or cells are not directly attached to the support surface, but simply trapped inside the polymer matrix. Entrapment is carried out by mixing the biocatalyst into a monomer solution, followed by polymerization initiated by a change in temperature or by a chemical reaction.

Polymers like polyacrylamide, collagen, cellulose acetate, calcium alginate or carrageenan etc are used as the matrices.

Advantages of entrapment:

- Loss of enzyme activity upon immobilization is minimized.

Disadvantages of entrapment:

- The enzyme can leak into the surrounding medium.
- Another problem is the mass transfer resistance to substrates and products.
- Substrate cannot diffuse deep into the gel matrix.

1. Occlusion within a cross linked gel:

In this entrapment method, a highly cross-linked gel is formed as a result of the polymerization which has a fine "wire mesh" structure and can more effectively hold smaller enzymes in its cages.

Amounts in excess of 1 g of enzyme per gram of gel or fibre may be entrapped.

Some synthetic polymers such as polyarylamide, polyvinylalcohol, etc. and natural polymer (starch) have been used to immobilize enzymes using this technique.

2. Microencapsulation:

This entrapment involves the formation of spherical particle called as "microcapsule" in which a liquid or suspension of biocatalyst is enclosed within a semi permeable polymeric membrane.

# Lipozyme

Lipozyme is a type of industrial enzymes and lipases are special class of Lipozyme. Lipases (triacylglycerol acyl hydrolases) are a class of enzymes, which catalyze the hydrolysis of long chain triglycerides. In humans and monogastric animal species, lipases enzyme naturally occur in the stomach and pancreas where they function to digest fats and lipids. For industrial applications, animal derived lipases are still used in some specific medical sectors (e.g. pig pancreas lipase to supplement lipase-deficient patients).

Microbial lipases are produced by fungal, yeast, and bacterial species. The following micro-organisms account for most of the industrial microbial lipases production: Candida sp., Aspergillus sp., Rhizomucor sp., Rhizopus sp., Humicola sp., Yarrowia lipolytica and Pseudomonas sp. They are developed, produced and marketed by the key enzyme players for different applications.

Industrial lipases application scope covers various industries and applications such as oleo-chemicals, detergents, polymers, food processing, pharmaceutical, waste, cosmetics and biodiesel.

## Lipases for the Food and Agro-industrial Applications

Lipases have many applications and benefits in the food and agroindustries, where they have quantitative and qualitative impacts. For instance, in the vegetal oil processing, lipases allow a significant increase in oil yield and at the same time the end product is of better appearance. In baking and dairy applications, lipases are added to enhance and accelerate the development of aromatic notes. In the food industry, lipases find a great interest in the egg processing sector to enhance the emulsification properties of egg yolk lipids.

Lipases are also being developed to create new functional ingredients and functional foods, such as cocoa butter equivalents or human milk fat equivalents.

The following lipase applications are reviewed: dairy, baking, cocoa butter substitutes, human milk fat substitutes, egg processing and edible oil production.

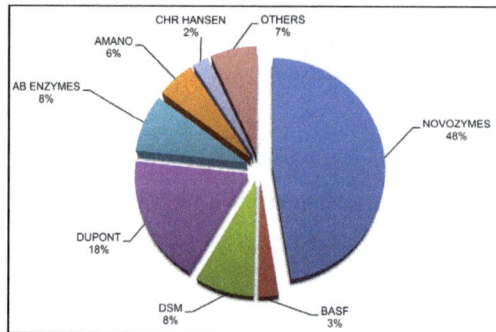

Industrial enzymes market share, estimation by company.

Table: Some examples of commercial lipases from the leading global enzyme suppliers.

| enzyme producer | Product examples | Application |
|---|---|---|
| NOVOZYMES | Lipase 435 | Multipurpose lipase |
| | Lipolase | Detergents |
| DSM | Gumzyme | Oil degumming |
| | Maxapal A2 | Egg processing |
| AB ENZYMES | Rohalase PL, Rohalase F, | Oil degumming |
| | Veron Hyperbake | Baking |
| AMANO | Lipase DF | Baking |

## Dairy

Lipases are used to break down milk fats and give characteristic flavors to cheeses. The flavor comes from the free fatty acids produced when milk fats are hydrolyzed. Both animal and microbial enzymes are used in the industry. Animal lipases are obtained from kid, calf and lamb, while microbial lipase is derived by fermentation, mainly with the fungal species Rhizomucor meihei. Animal and microbial lipases have different action patterns and the industry uses both according to the expected flavor profile (hydrolysis of the shorter fats is preferred because it results in the desirable taste of many cheeses while the hydrolysis of the longer chain fatty acids can result in "soapy" notes).

## Baking

In baking industry, (phospho) lipases can be used to substitute or supplement traditional emulsifiers through the degradation of wheat lipids to produce emulsifying lipids in situ. Lipase in baking also enhance the flavor of bakery products by liberating short-chain fatty acids through esterification. In synergy with other commonly used baking enzymes (amylases, xylanases), lipases contribute to increase the loaf volume and improve crumb firmness, allow to prolong the shelf-life of baked products and improve their texture and softness.

### Cocoa Butter Substitutes

Cocoa butter is a fat mainly formed by triacylglycerides (TAGs) with two saturated fatty acids (palmitic and stearic acids) and one mono-unsaturated fatty acid (oleic acid). Thanks to its unique structure, cocoa butter shows unique properties in the food processing industry (mouthfeel, melting behaviour). However, cocoa butter supply is uncertain and its price highly fluctuant. Enzymatic processes have been developed to catalyse interesterification reaction of different edible oils (such as sunflower oils) to produce fats having a composition and properties comparable to those of cocoa butter. Such products are called cocoa butter equivalents or cocoa butter substitutes (CBS).

### Human Milk Fat Substitutes

Human milk fat (HMF) contains different lipids: oleic (30– 35%), palmitic (20–30%), linoleic (7–14%) and stearic acids (5.7–8%). Unlike in vegetable oils and in cow's milk fat, in HMF, palmitic acid, the major saturated fatty acid, is mostly esterified at the sn-2 position of the TAGs, while unsaturated fatty acids are at the external positions. The fatty acid profile of HMF has a crucial effect on its digestibility and intestinal absorption in infants. Human milk fat substitutes (HMFS) have been obtained by sn-1,3 lipase-catalyzed acidolysis of tripalmitin, butterfat, palm oil, palm stearin or lard (rich in palmitic acid in sn-2 position) with free fatty acids (FFA) from different sources.

| Industry sector | Applications and benefits |
|---|---|
| Fats and oleo-chemistry | The current trend in the oleo-chemical industry is to involve the use of immobilized lipase to catalyze the hydrolysis, esterification and inter-esterification of oils and fats as an alternative to physico-chemical processes (energy saving, specificity of reactions) |
| Detergents | Next to proteases as detergent additives, lipases are the second most important group of detergent enzymes to contribute to oil and fats stain and traces removal. Lipases are used in both laundry and dishwashing formulations commercial detergents where they have been optimized to operate under various pH and temperature conditions. Lipex and Lipolase from Novozymes are two examples of lipases sold to the detergent industry. |
| Polymers | Lipases can be used to assist the production of polymers, such as polyester, PLA (polylactide), PCL (polycaprolactone) an many others, in alternative to chemically catalysed reactions Immobilised lipases are also used in lignin transesterification reactions to produce lignine oleate, bringing new properties to the polymer. |
| Food processing | Lipases are utilized in dairy products for flavor development but also in the processing of other foods such as meat products, baked foods, cocoa butter processing and others. |
| Medical and pharmaceutical sectors | Due to their high level of specificity, lipases can be used to produce active pharmaceutical compounds. For example, enantioselective enzymes are used as an alternative technology to chiral chromatography Lipases are also applied in the production of lysophospholipids from phospholipids and to extract and produce functional lipids. |
| Pulp and paper | The presence of hydrophobic components (mainly triglycerides and waxes) in wood are detrimental to many of the processes in the production of paper and pulp, and lipases can be used to remove those undesirable triglycerides. |
| Waste / effluent / sewage treatment | Lipases are added to eliminate the thin layers of fats formed at the surface of waste water reservoirs and therefore recover active oxygen transport conditions which are necessary in maintaining optimum biomass growth. |
| Cosmetics and perfumery | Lipases are used in the productin of surfactants (e.g. mono-acylglycerols and diacylglycerols) via the controlled esterification of glycerols In the cosmetic and fragrances industries lipases can be used in the synthesis of citronellyl butyrate andvalerate. |
| Biodiesel | Lipases are used to produce biodiesel from various feedstocks such as palm oil or animal fats. Thermostable lipases have been developed to optimize the application of enzymes in biodiesel production. |

# Egg Processing

Eggs provide functional ingredients to the food industry with a variety of properties including foaming, gelation, emulsifying in batters and mayonnaise and improved texture of baked goods. Egg lipids are responsible for the emulsifying properties. Lipases can greatly improve egg lipids emulsifying power for better performance and lower egg yolk addition rate in processed food recipes, such as dressings and mayonnaise-like products.

Egg yolk emulsification for dressings- the global production of emulsified dressings is estimated at 3 millions of metric tons per year, consuming roughly 150 000 metric tons of egg yolk in their process. One third of the market for emulsified dressings is concentrated in Russia and Eastern Europe countries. The market is highly industrialized, with global players such as Nestlé, Kraft and Unilever. Egg yolk is a complex oil-water emulsion composed of 50% water, 32% lipids and 16% protein. Approximately, 1/3 of the lipids are phospholipids, of which approximately 80% is phoshatidylcholine (PC). Egg yolk also contains phosphatidylethanolamine.

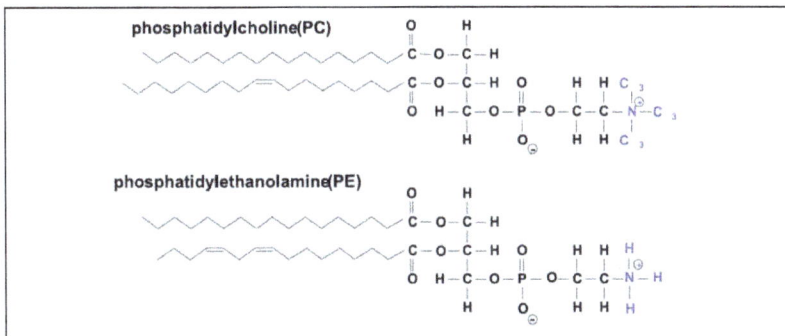

Chemical structure of most important phospholipids in egg yolk

The enzymatic conversion of egg yolk phospholipids into lyso-phospholipids will increase the emulsion stability. Enzyme producers have developed different phospholipases (PL) acting at different positions.

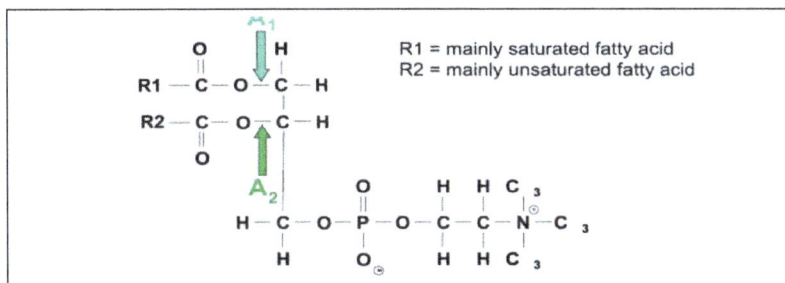

Action of phospholipases in $A_1$ and $A_2$ positions

Egg processing lipases can be extracted from pork pancreas or from micro-organisms (e.g. Maxapal $A_2$, DSM, NL). Using pork pancreas lipases, the conversion rate from phospholipids to lysophospholipids is typically over 80% after 1 hour reaction. However the application of such enzyme is limited by its animal origin (vegan consumers, Halal/Kosher restrictions). Lipases cloned and expressed in the fungus Aspergillus niger (such as DSM Maxapal $A_2$) also reach high conversion yield but some industry players are not in favour of using enzymes produced by such modified micro-organisms (MGM).

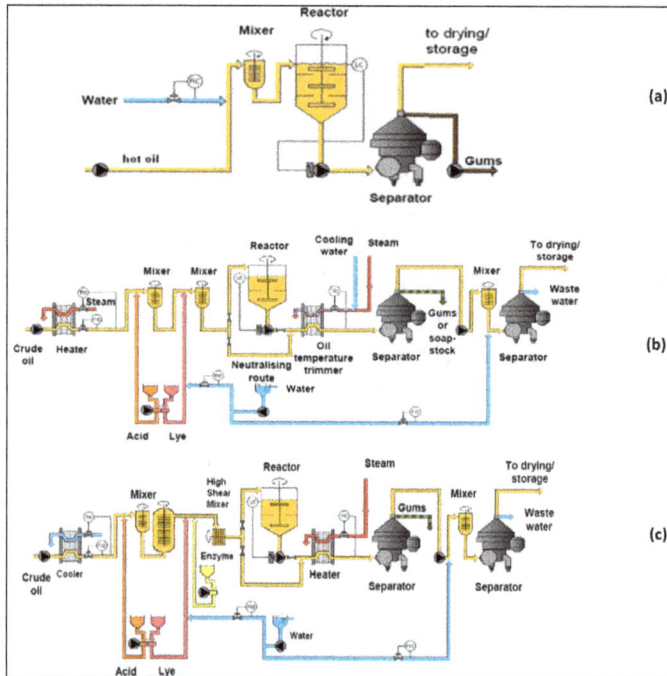

Oil degumming flowcharts comparing three technologies (a) high temperature water extraction (b) process with acidification and (c) enzyme assisted process (from Alfa-Laval technical documentation).

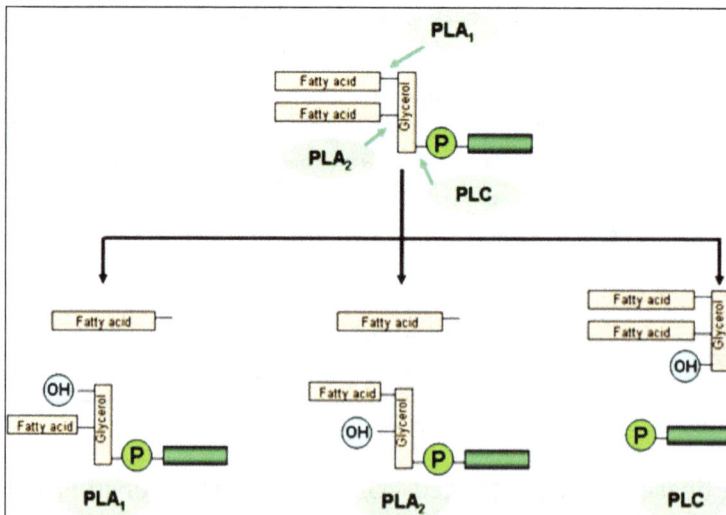

Mode of action of different phospholipases (PL) used for oil degumming.

# Edible Oil Production

Edible oils from plants are used in different industries: food, feed and fuel. Food consumption of vegetal oils is close to 150 million of metric tons per year, of which 1/3 being palm oil. Production of edible vegetal oils is dominated by three leading global players: ADM, Bunge and Cargill. When refining vegetal oils, it is necessary to remove impurities that affect the yield and moreover the taste, smell, visual appearance, and storage stability of the oil. The most important class of impurities are phopholipids, often referred as "gums". For decades, the industry has been using thermochemical processes to remove those phospholipids (PL) from crude oil, and more recently different enzymes solutions (based on phospholipases) have been proposed by the enzyme companies. The hydrolysis of the PL by phospholipases at the water-oil interface will inhibit the emulsion and release the oil trapped in the gum phase. Flowcharts of non-enzyme assisted processes vs. enzyme assisted process are presented in Figure. Enzymatic degumming can be applied to oils from rapeseed/canola, soyabean, ricebran, corn, sunflower seeds, and palm with the following benefits:

- Higher oil yields (up to 2% increase);

- Energy savings (operates at 50–60 °C vs. 85 °C for conventional process);

- Very low oil losses;

- Easier pumping and separation operations;

- Limited formation of soaps;

- Lower water consumption.

Three main phospholipases (PL) used by the industry are categorized according to their fatty acid-glycerol cleavage pattern and are referred as $PLA_1$, $PLA_2$ and PLC. For optimum performance, enzymatic oil degumming can also involve the use of PL combinations, either in blends or in sequential additions. It is important to note that besides their different cleavage patterns, each PL developed by the enzyme industry has unique properties in terms of pH and temperature profiles according to their origin and mode of production. PL used in the oil industry are from microbial origin (Trichoderma, Aspergillus, Pichia) or still from porcine pancreas. Leading suppliers for the PL range of enzymes are Novozymes, DSM, Dupont, AB Enzymes.

As the oil industry is implementing the enzyme assisted degumming, PL product ranges from the leading suppliers are in constant evolution. Product innovation is driven by the need for enzymes with improved performance, higher temperature resistance and ability to operate at lower pH. For instance, Novozymes introduced in 2000 the Lecitase Ultra 10L (pancreatic $PLA_2$) and the Lecitase (microbial $PLA_1$), followed in 2003 by Lecitase Ultra (thermostable microbial $PLA_1$) and recently announced the launch of a new acid tolerant and thermostable PLA1 from Talaromyces leycettanus.

# Phytase

Phytases (myo-inositol hexakisphosphate phosphohydrolases) represent a subgroup of phosphatases that are capable of initiating the stepwise dephosphorylation of phytate, the major storage form of phosphate in plant seeds. For this reason, it is used in several food processing and preparation methods in order to reduce the phytate content of cereals and legumes. This class of enzymes has also been found increasingly interesting for their use in processing and manufacturing of food for human consumption, particularly because the decline in food phytate results in an enhancement of mineral bioavailability. It is considered a way to reduce the risk of mineral deficiency in vulnerable groups including childbearing women, strict vegetarians, babies consuming soy-based infant formulas, and the inhabitants of developing countries. There is, however, growing evidence to demonstrate the beneficial role played by phytic acid in all human organisms.

Some ingredients possess intrinsic phytase activity, which varies greatly among plant species. Corn and soybean meal contain negligible levels of phytase activity compared to wheat, which contains considerably higher levels of intrinsic phytase. The majority of phytase activity in cereal grains is found in the aleurone layers.However, this may be lost when ingredients are subjected to high temperatures, such as during the pelleting process. Commercially available exogenous phytases are commonly derived from either fungi or bacteria, such as Aspergillus niger and Escherichia coli,6 but can also be expressed in yeasts.

## Affects the Efficacy of Phytase

Several factors can influence the efficacy of phytase, including the amount of phytate in the diet, the amount of phytase added to the diet, and the type of phytase. Phytase derived from E coli bacteria is also more efficacious than the fungal phytases in terms of the amount of phosphorus released per unit of phytase, according to published data. However, analytical techniques being used to determine phosphorus release vary among commercial phytase manufacturers. Because of this, the amount of phosphorus released per unit of phytase may differ between two phytase products, as shown in a recent study using a standard assay procedure.Thus, depending on the assay used, different results of phytase activity may be reported.

Because phytase is a protein, it is susceptible to denaturation when subjected to excessive heat, such as during pelleting. This may be addressed by spraying liquid phytase onto the cooled pellets to maintain the stability of the enzyme. In addition, heat-stable phytases are available. Phytase is also sensitive to degradation when stored in premixes under high temperature and moisture conditions. Hence, proper storage procedures and frequent rotation of products containing phytase must be practiced. Phytase products should be stored only in cool, dark, dry areas. The manufacturer's recommendations should always be followed, especially when phytase is included in vitamins and trace-mineral premixes.

# References

- Chemical-technology-pep-industrial-enzymes: ihsmarkit.com, Retrieved 1 May, 2019

- Industrial-enzyme, chemistry: sciencedirect.com, Retrieved 18 January, 2019

- Taherzadeh, Madhavan; Nampoothiri, Christian (2015). Industrial Biorefineries and White Biotechnology. Elsevier B.V. ISBN 978-0-444-63453-5

- Immobilized-enzymes: nature.com, Retrieved 19 June, 2019

- Enzyme-immobilization: enzymeimmobilization.blogspot.com, Retrieved 22 March, 2019

- Phytase, neuroscience: sciencedirect.com, Retrieved 30 July, 2019

# Permissions

# Index